Cambridge IGCSE Mathematics

Revision Book

By Vali Nasser

Copyright © 2017

E-book editions are also available for this title. For more information email: valinasser@gmail.com

All rights reserved by the author. No part of this publication can be reproduced, stored in a retrieval system, or transmitted in any form or by any means, electronic, mechanical, photocopying, recording or otherwise, without the prior permission of the publisher and/or author.

ISBN-13: 978-1975799243

ISBN-10: 1975799240

First Edition: August 2017

Every effort has been made by the author to ensure that the material in this book is up to date and in line with the requirements to pass the Cambridge IGCSE Mathematics at the time of publication. The author will also do his best to review, revise and update this material periodically as necessary. However, neither the author nor the publisher can accept responsibility for loss or damage resulting from the material in this book

INTRODUCTION	7
NUMBERS:	9
Standard Index form	12
INDICES, RATIONAL, IRRATIONAL NUMBERS AND SURDS	14
Fractional Indices	14
Rational and Irrational Numbers	15
Converting recurring decimals into fractions	15
ROUNDING AND ESTIMATING	18
UPPER AND LOWER BOUNDS	20
FRACTIONS, DECIMALS AND PERCENTAGES	21
Arithmetic questions involving percentages and fractions	23
Compound Interest, Appreciation and Depreciation	25
SIMPLIFYING AND MANIPULATING FRACTIONS	27
Finding fraction of an amount	28
Adding and Subtracting Fractions	28
Adding and subtracting mixed numbers	31
Multiplying Fractions	32
Division of Fractions	32
Multiplying mixed numbers together	33

Dividing mixed numbers together	33

PROPORTIONS AND RATIO — 35

Scales and ratios	35
Conversions	36

SETS AND VENN DIAGRAMS — 39

Symbols and notation associated with Sets	40

PRACTICE QUESTIONS 1 - NO CALCULATORS ALLOWED — 48

Answers to Non-Calculator Practice Questions	50

PRACTICE QUESTIONS 2: CALCULATORS ALLOWED — 51

Answers to Calculator Practice Questions	52

BASIC ALGEBRA — 53

Multiplying positive and negative numbers.	54
Dividing positive and negative numbers.	54
Simplifying algebraic expressions	55

SIMPLIFYING ALGEBRAIC FRACTIONS — 58

ALGEBRAIC SUBSTITUTION AND FORMULA — 59

SOLVING EQUATIONS — 63

Linear Equations	63

SIMULTANEOUS EQUATIONS — 66

SOLVING WORD PROBLEMS USING ALGEBRA — 68

SOLVING QUADRATIC EQUATIONS — 70

SOLVING CUBIC EQUATIONS — 75

Using Trial and Improvement — 75

CHANGE THE SUBJECT OF A FORMULA — 76

PROPORTIONALITY — 78

PRACTICE QUESTIONS 3 -NO CALCULATORS ALLOWED — 81

Answers to non-calculator practice questions: — 82

PRACTICE QUESTIONS 4: CALCULATORS ALLOWED — 83

Answers to calculator based practice questions — 84

TRANSFORMATION GEOMETRY — 85

Bearings — 93

AREAS AND VOLUMES OF COMMON SHAPES — 94

Perimeters, Areas and Volumes of common shapes — 94

LOCI — 97

CIRCLE THEOREM — 100

LINEAR EQUATIONS — 104

GRAPHS — 107

SOLVING EQUATIONS USING GRAPHICAL METHODS	109
INVERSE AND COMPOSITE FUNCTIONS	112
PLAN AND ELEVATION	114
TRIGONOMETRY FOR RIGHT ANGLED TRIANGLES	115
TRIGONOMETRY FOR NON- RIGHT ANGLED TRIANGLES	117
GRAPHS OF TRIGONOMETRICAL FUNCTIONS	119
TRIG IDENTITIES	122
SIMILARITY AND CONGRUENCE	123
PYTHAGORAS' THEOREM	127
VOLUMES AND SURFACE AREAS	129
VECTORS	133
ARITHMETIC SEQUENCES	136
MATRICES	138
MULTIPLYING A MATRIX BY ANOTHER MATRIX	140
PRACTICE QUESTIONS ON MATRICES	145
ANSWERS TO PRACTICE QUESTIONS ON MATRICES	146
STATISTICS	148

SCATTER GRAPHS 156

BOX AND WHISKER PLOTS 161

CUMULATIVE FREQUENCY DIAGRAMS 163

PROBABILITY 168

DENSITY MASS AND VOLUME 175

PRACTICE QUESTIONS 5 – NO CALCULATORS ALLOWED 176

Answers to Practice Questions (non-calculator section) 178

PRACTICE QUESTIONS 6: CALCULATORS ALLOWED 180

Answers: Practice – Calculator Paper 182

Introduction

This book has been specially written to help you revise for IGCSE Mathematics for the Cambridge Examination Board.

Although the book starts gently with some basic reminders its main aim is to work towards getting a grade B, A or A* in the exam.

The topics you need to be familiar with are number, algebra, graphs, geometry, mensuration, co-ordinate geometry, trigonometry, functions, inverse functions, composite functions, matrices and transformations.

Hopefully, some of the material will already be familiar to you. If you work through most of the examples in this book as well as the practice questions you are very likely to pass with a very good grade. Sometimes there is more than one way of working out a given problem. If you feel comfortable with another method go along with that. However, it is important to show your working. You will only get minimal marks if you don't show your working in full.

This book has many worked examples and practice questions to help you become familiar with what to expect in the exam.

Good luck with your exams.

About the Author

The author of this book has experience in both consultancy work and teaching. He graduated initially from the University of St. Andrews in Mathematics and Physics. He was also project manager at OCR working in conjunction with the Teaching Agency in implementing the initial phase of the Numeracy Skills Testing project for Teachers. As a specialist mathematics teacher he has subsequently tutored and taught mathematics in schools as well as in adult education. The author's initial book 'Speed Mathematics Using the Vedic System' has a significant following and has been translated into Japanese and Chinese as well as German. His book 'Pass the QTS Numeracy Skills Test with Ease' has been very popular with teacher trainees and 'Pass the Numerical Reasoning Tests' is very popular with graduates wanting to further their career. He hopes that this IGCSE revision book for Cambridge Exam board will help students to get a good grade in this subject.

Numbers:

Integers: These are whole numbers that include both positive and negative numbers including 0. So for example ....-5,-4,-3,-2, 0, 1, 2, 3, 4, ... are all integers.

Multiples: These are simply numbers in the multiplication tables.

For example the multiples of 6 are 6, 12, 18, 24, 30, ...

Factors: A factor is a number that divides exactly into another number for example, the number 2 in the case of even numbers.

3 is a factor of 9, as 3 goes exactly into 9. Other factors of 9 are 1 and 9.

15, has two factors other than 15 and 1. The two factors are 5 and 3, since both these numbers go exactly into 15. **Example:** Find all the factors of 21. The factors are: 1, 3, 7 and 21 (since all these numbers divide exactly into 21)

Prime numbers: A prime number is a natural number that can be divided only by itself and by 1 (without a remainder). For example, 11 can be divided only by 1 and by 11. Prime numbers are whole numbers greater than 1. So for example the first 10 prime numbers are: 2, 3, 5, 7, 11, 13, 17, 19, 23 and 29. **Be careful that an odd number is not necessarily a prime number.** For example **9 is not a prime number** as its factors are 1, 3 and 9 and **prime numbers should have only two factors, 1 and the number itself. Also, note that 2 is a prime number, the only even number that can be divided by 1 and itself!**

Prime Factors: Some numbers can be written as a product of prime factors.

Example1: Write 28 as a product of prime factors.

Dividing 28 by the first prime factor 2 we are left with 14. Dividing 14 again by the first prime factor 2, we get 7. Now we can no longer divide 7 by the first prime factor 2. The next possible prime factor for 7 is obviously 7. **Hence 28 can be written as 2×2×7 or $2^2 \times 7$**

Example 2: Write the number 300 as a product of prime factors.

Step1: Divide by 300 by 2 to get 150,

Step 2: Divide 150 by 2 to get 75

Step 3: Divide 75 by 3 to get 25, Step 4: Divide 25 by 5 to get 5

Step 4: Divide 5 by 5 to get 1. Hence, **300 = 2×2×3×5×5 or $2^2 \times 3 \times 5^2$**

Another method of finding prime factors: Break down the required number at the top by dividing by prime factors starting with the lowest prime factor as shown below:

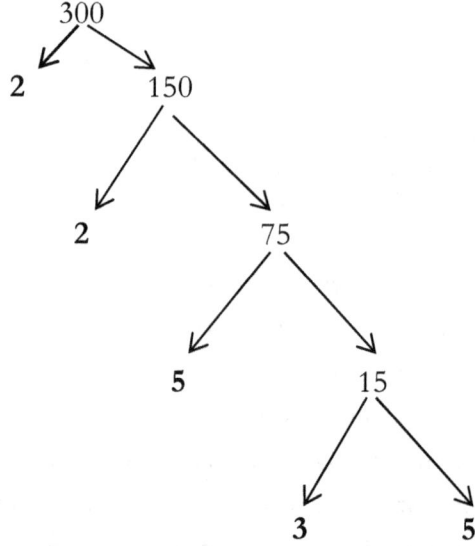

Hence the product of prime factors for **300 = 2×2×3×5×5 or $2^2 \times 3 \times 5^2$**

Lowest Common Multiple (LCM)

This is essentially the smallest number that will divide exactly by the numbers given. Consider the examples below:

Example: Find the LCM of 15 and 45

One method is to find the multiples of both numbers and identify the lowest common multiple as shown below:

Multiples of 15 = 15, 30, **45**, 60, 90, ……..

Multiples of 45 = **45**, 90, 135, 180, ……

Clearly **45** (the highlighted number above) is the smallest number that is divisible by 15 and 45.

Highest Common Factor (HCF)

This is the biggest number that will divide exactly into all the numbers given

Example: Find the HCF of 8 and 32.

First find the factors of each number given

Factors of 8 = {1, 2, 4, **8**}, Factors of 32 = {1, 2, 4, **8**, 16, 32}

You can see that the number 8 is the highest **common** factor which divides into 8 and 32 exactly.

Square numbers and square roots

Squaring a number is simply multiplying a number by itself.

So 4^2 means $4 \times 4 = 16$, 12^2 means $12 \times 12 = 144$ and so on.

The square root is written like this $\sqrt{}$ and means finding a number which when multiplied by itself gives you the number inside the square root.

Example: Find $\sqrt{16}$. The answer is clearly 4. Since $4 \times 4 = 16$

Let us consider some other square roots.

$\sqrt{49} = 7$, $\sqrt{121} = 11$, $\sqrt{100} = 10$, $\sqrt{225} = 15$,

$\sqrt{256} = 16$, $\sqrt{324} = 18$

Cubes and cubic roots

Cubing a number is simply multiplying the number by itself three consecutive times. A cube of a number is written as x^3, where x is the number.

So, for example, 5^3 means $5 \times 5 \times 5 = 25 \times 5 = 125$

Similarly, $6^3 = 6 \times 6 \times 6 = 216$, $7^3 = 7 \times 7 \times 7 = 343$, $9^3 = 9 \times 9 \times 9 = 729$, $10^3 = 10 \times 10 \times 10 = 1000$

Cube Roots

Cube roots are found by finding a number which when cubed gives you the number inside the cube root.

So for example the cube root of 125 is written as $\sqrt[3]{125}$

Also we know that $5 \times 5 \times 5 = 125$, so that $\sqrt[3]{125} = 5$

Standard Index form

Standard Index form or scientific notation helps us to write **very large** or **very small** numbers in a more elegant way.

Example 1: Write three million in standard index form.

Three million = 3000,000 = 3 × 1000,000

Now 3 × 1000,000 = 3 × 10^6

3 × 10^6 is the required standard index form.

Example 2: Write 4000,000 in standard index form

We know that 4000,000 = 4 × 1000,000

Hence we can write 4 × 1000,000 as 4×10^6. This is in standard index form.

Example 3: Change 4500,000 to standard index form

We know that 4500,000 = 4.5 × 1000,000

So 4.5 × 1000,000 = 4.5×10^6. This is in standard index form.

Now consider very small numbers:

Example 4: Write 0.0004 in standard index form

We can re-write 0.0004 as 4 ÷ 10,000

Now, 4 ÷ 10,000 can be written as $\dfrac{4}{10 \times 10 \times 10 \times 10} = 4 \times 10^{-4}$

Notice, $\dfrac{1}{10} = 10^{-1}$, $\dfrac{1}{100} = 10^{-2}$, $\dfrac{1}{1000} = 10^{-3}$ and $\dfrac{1}{10000} = 10^{-4}$

Summary:

To convert a number to standard index form change the number to the form a × 10^n where 1 ≤ a < 10 (a is between 1 and 10) and n is a whole positive or negative number)

Indices, Rational, Irrational Numbers and Surds

You should now be familiar with squares, square roots, cubes and cube roots. Powers, Indices/Index Numbers or exponents are simply the power by which a base number is raised. So just as 4^3 (4 cubed) means 4 to the power of 3, these 'powers' as mentioned earlier are also referred to as indices or index numbers. So 5^6 simply means 5 raised to the power of six. So in this case 5^6 means 5×5 5×5×5×5! (5 is called the base number and 6 is the power or index number) It is interesting to note that if you multiply two or more same base numbers with indices for example: $5^6 \times 5^3$ you simply add the indices to get 5^9 (5 to the power 9). Reason: 5^6 means 5×5×5×5×5×5 and 5^3 means 5×5×5 so $5^6 \times 5^3$ = (5×5×5×5×5×5) × (5×5×5)= 5^9

Similarly, for division, you simply subtract the indices. Consider $5^6 \div 5^3$. This means we need to work out $\frac{5 \times 5 \times 5 \times 5 \times 5 \times 5}{5 \times 5 \times 5}$ which cancels down to 5×5×5 or 5^3 So you can see that when dividing the **same** base numbers with indices you simply subtract the indices.

The examples below will help you to consolidate the manipulation of the same base numbers with indices.

Example 1: $7^8 \times 7^4 \times 7^6 = 7^{18}$ (simply add the indices 8 + 4 +6 =18, hence the answer is: 7^{18})

Example 2: $9^{12} \div 9^5 = 9^7$ (simply subtract 5 from 12 to get 7, hence the answer is: 9^7)

Finally, you can also have negative indices which are inverses of the base numbers with the appropriate indices.

Example 1: $5^{-1} = \frac{1}{5}$ (Also called the reciprocal of 5).

Example 2: $6^{-2} = \frac{1}{6^2}$

Fractional Indices: Examples: (i) $2^{1/2}$ (2 to the power of $\frac{1}{2}$), (ii) $27^{1/3}$ (27 to the power of $\frac{1}{3}$. (It's worth noting that $2^{1/2}$ is the same $\sqrt{2}$

Rational and Irrational Numbers

Numbers can be either rational or irrational

Any number that can be written as p/q is a rational number, where p and q are whole numbers and q is not zero. Basically, the number is well defined and we know or can predict its pattern.

Examples of rational numbers are: $5 = \frac{5}{1}$, $-2 = \frac{-2}{1}$, $\frac{1}{2} = 0.5$, $\frac{2}{5} = 0.4$, $\frac{1}{3} = 0.33333$ **(recurring)**

$\frac{0}{5} = 0$, $\frac{4}{33} = 0.1212121212.....$

Examples of irrational numbers are: $\pi, \sqrt{2}, \sqrt{3}$ or $5\sqrt{7}$

For square roots and cube roots those with perfect roots are rational whereas others are irrational. So for example $\sqrt{25} = 5$ is rational, $\sqrt[3]{27} = 3$ is rational but as we saw earlier $\sqrt{2}$ is irrational.

For example π or $\sqrt{2}$ do not have a predictable pattern. We can approximate them but not calculate them exactly.

Converting recurring decimals into fractions

Example1: Convert the recurring decimal 0.5555... to a fraction

Let $r = 0.\dot{5}$

Multiply the both sides by 10 to get, **10r = 5.$\dot{5}$**

Now subtract the original from **10r = 5.$\dot{5}$** to get **9r = 5**

Hence $r = \dfrac{5}{9}$

Example2: Convert $0.2\dot{1}$ to a fraction

let $r = 0.\dot{2}\dot{1}$, multiply the both sides by 100 to get, 100r or $100r = 21.\dot{2}\dot{1}$

Now subtract to get **99r = 21.**

Which means $r = \dfrac{21}{99} = \dfrac{7}{33}$

Summary for Indices:

Rules of indices: (1) $a^m \times a^n = a^{m+n}$

(2) $a^m \div a^n = a^{m-n}$

(3) $(a^m)^n = a^{m \times n}$

(4) $a^0 = 1$

(5) $a^{-1} = \dfrac{1}{a}$

(6) $a^{-m} = \dfrac{1}{a^m}$

Surds

Surds are simply expressions with irrational square roots. There are some useful rules associated with them.

1: $\sqrt{2} \times \sqrt{2} = \sqrt{4} = 2$

2: $\sqrt{3} \times \sqrt{2} = \sqrt{6}$

3: $\dfrac{\sqrt{6}}{\sqrt{2}} = \sqrt{\dfrac{6}{2}} = \sqrt{3}$

4: $(\sqrt{p} + \sqrt{q})^2 = (\sqrt{p} + \sqrt{q}) \times (\sqrt{p} + \sqrt{q}) = p + 2\sqrt{pq} + q$

5. $(p + \sqrt{q})(p - \sqrt{q}) = p^2 + p\sqrt{q} - p\sqrt{q} - q = p^2 - q$

6. $\dfrac{2}{\sqrt{3}}$ (Multiply top and bottom by $\sqrt{3}$) so we have: $\dfrac{2}{\sqrt{3}} \times \dfrac{\sqrt{3}}{\sqrt{3}} = \dfrac{2\sqrt{3}}{3}$

we call this rationalising the denominator.

7. $\dfrac{2}{1-\sqrt{3}}$ to simplify this we need to <u>rationalise</u> the denominator.

(to do this we simply multiply top and bottom by $(1 + \sqrt{3})$

So we have, $\dfrac{2}{1-\sqrt{3}} = \dfrac{2}{1-\sqrt{3}} \times \dfrac{1+\sqrt{3}}{1+\sqrt{3}} = \dfrac{2(1+\sqrt{3})}{1-3} = \dfrac{2(1+\sqrt{3})}{-2} = -(1 + \sqrt{3})$

Note: <u>Leaving your answers as surds is quite respectable. Since you can't work out the exact answer on a calculator!</u>

Rounding and estimating

We will start simply with rounding numbers to the nearest 10 and 100

Consider the number 271

Rounded to the nearest 10 this number is 270

Rounded to the nearest 100 this number is 300

The principle is that if the right hand digit is lower than 5 you drop this number and replace it by 0. Conversely, if the number is 5 or more drop that digit and add 1 to the left

Try a few more:

5382 to the nearest 10 is 5380

5382 to the nearest hundred is 5400

5382 to the nearest 1000 is 5000

This rule can also be applied to decimal numbers:

3.7653 rounded to the nearest thousandth is 3.765

3.7653 rounded to the nearest hundredth is 3.77

3.7653 rounded to the nearest tenth is 3.8

3.7653 rounded to the nearest unit is 4

Tip: remember to use common sense when rounding in real life situations.

Example: A teacher wants to keep 120 English text books in the same size boxes. She can fit 23 text books in a box. How many boxes will she need?

Method: Number of boxes required will be 120÷23= 5.2 (to one decimal place). But clearly, she cannot have 5.2 boxes. So she needs to have 6 boxes

Estimating calculations quickly

Example 1: Work out $(2.2 \times 7.12)/4.12$

We can quickly estimate that this is roughly equal to $(2 \times 7)/4$

$=14/4$ which is around 3.5 or 4 rounded to the nearest unit.

The actual answer is: 3.8 (to 1 decimal place)

Example 2: Estimate the value of $\sqrt{15.9} \div 1.99^2$

Now $\sqrt{15.9}$ is approximately $\sqrt{16} = 4$,

Likewise 1.99 is approximately 2, hence $2^2 = 2 \times 2 = 4$

So $\sqrt{15.9} \div 1.99^2$ is approximately $= 4 \div 4 = 1$

Now consider Significant Figures (S.F.)

Normally, the first digit is the first significant figure except when it is 0, when you do not count it. Remember to get the size of the number right when working out significant figures.

Example 1: Write 53.6 to 1 s.f. Answer is 50 (It is not 5)

Example 2: Write 262.7 to 2 s.f. Answer is 260 (It is not 26)

Example 3: Write 0.0384 to 1 s.f. Answer is 0.04 (ignore the 0 at the beginning)

Upper and Lower Bounds

Example 1: A table is 8 metres long. Find the upper and lower bounds.

Method: The length is 0.5m on either side. This means the upper bound is 8.5m and the lower bound is 7.5m. However we can write this as 8.5m ≤ length < 9m

Example 2: A rectangle is 5cm by 4cm. Find the maximum and minimum possible areas.

Method: To find the maximum area first work out the maximum possible values of the sides of the rectangle. Hence the maximum area is 5.5 × 4.5 =**24.75cm^2** and similarly the minimum area is 4.5×3.5 = **15.75cm^2**

Example 3: A circle's circumference is measured to the nearest 0.1 cm assuming the circumference is 15.5cm. What is the maximum and minimum value of the circumference?

This is straight forward – the maximum value is **15.55cm** and the minimum value is **15.45cm**.

Example 4: x = 3.6 and y = 2.1 both given to 1 decimal place. What are the maximum and minimum values of x – y?

Method: The maximum value of x is 3.65 and the minimum 3.55. Similarly the maximum value of y = 2.15 and the minimum 2.05. **Hence maximum 3.65 – 2.05 = 1.60** and the **minimum value is 3.55 – 2.15 = 1.40**

Fractions, decimals and percentages

I am sure most of you are aware that $\frac{1}{2}=0.5$. This in turn is equal to 50%.

It is worth reviewing this fact. In addition, you should try and remember the following other equivalences if you have forgotten them:

Fractions, decimals and percentage equivalents

Fractions	Decimal	Percentage
$\frac{1}{2}$	0.5	50%
$\frac{1}{4}$	0.25	25%
$\frac{3}{4}$	0.75	75%
$\frac{1}{10}$	0.1	10%
$\frac{1}{5}$	0.2	20%

If, we know $\frac{1}{2}$ = 0.5

We can deduce that $\frac{1}{4} = 0.25$

(Since a quarter is half of half)

Similarly $\frac{1}{8}$ is **0.125**

We can do this quickly because all we do is halve each decimal value.

Half of 0.5 is 0.25, Half of 0.25 is 0.125

We can of course continue this process.

Further if we know $\frac{1}{10} = 0.1$ **we can now work out** $\frac{2}{10}, \frac{3}{10}, \frac{7}{10}$ etc.

$\frac{2}{10} = 0.2$ (2 X 0.1), $\frac{3}{10} = 0.3$ (3 X 0.1), $\frac{7}{10} = 0.7$ (7 X 0.1), $\frac{9}{10} = 0.9$ (9 X 0.1)

Another useful fraction and decimal equivalent to remember is $\frac{1}{3}$ =0.333… (0.3 recurring)

The key equivalent percentages to remember are as follows:

$\frac{3}{4} = 75\%$, $\frac{1}{2} = 50\%$, $\frac{1}{4} = 25\%$, $\frac{1}{8} = 12.5\%$, $\frac{1}{10} = 10\%$

See summary box below

Summary:

Remember the following equivalences

$\frac{1}{2} = 0.5 = 50\%$, $\frac{1}{4} = 0.25 = 25\%$, $\frac{3}{4} = 0.75 = 75\%$, $\frac{1}{10} = 0.1 = 10\%$

Also if you can try to remember, $\frac{1}{5}$ =0.2 = 20%, and $\frac{2}{5}$ =0.4 =40%, $\frac{1}{3}$ =0.333… (0.3 recurring) =33.33% (to 2 decimal places)

To convert a fraction into a percentage, simply multiply the fraction by 100

Arithmetic questions involving percentages and fractions

Example 1: In a class of 25 pupils there are 12 girls and the rest are boys.

(1) What fraction consists of boys? (2) What percentage is this?

Method:

(1) Since there are 12 girls, there are 13 boys out of 25. So the fraction of boys is $\frac{13}{25}$

(2) The percentage of boys is $\frac{13}{25} \times 100 = $ **52%,** (Divide 100 by 25 to get 4. Then multiply 13 by 4 to get 52%)

Example 2: 30% of the pupils in a class are boys. There are 30 pupils altogether. How many pupils are girls in this class?

Method: If 30% of the pupils in a class are boys, clearly 70% are girls. So we need to find 70% of 30 pupils. Since 10% of 30 is 3, this means 70% corresponds to 3×7 = 21 girls. Hence, this class has 21 pupils that are girls.

Working out increase or decrease in percentages from original value. Example1: In a certain school 16 pupils got a grade B in Science in 2010. In the same school 20 pupils got a grade B in Science in 2011. What was the percentage increase in grade B's from 2010 to 2011 in this subject?

Method: Increase in number of grade B's = 20 – 16 =4. Original number of pupils =16. The increase of 4 was based on 16 pupils. To work out the percentage increase we simply divide the increase by the original number of pupils with grade B and multiply this by 100. That is $\frac{4}{16} \times 100 = \frac{1}{4} \times 100 = 25\%$

To work out decrease in percentages (uses the same principle as above)

Example 2: The original price of a classroom projector was: £150, the new price is reduced to £135. What is the percentage decrease in price? The decrease in price is £150 - £135 = £15. The decrease over the original price is $\frac{15}{150}$, to turn this into a percentage we multiply $\frac{15}{150} \times 100 = \frac{1500}{150} = 10\%$. So the decrease in percentage price is 10%. The basic formula to work out increase or decrease percentage change is shown below:

$$\frac{difference\ between\ final\ and\ original\ value}{original\ value} \times 100$$

Miscellaneous questions involving fractions and percentages

Example 1: Finding fraction of an amount

Find $\frac{3}{4}$ of £600

First find $\frac{1}{2}$ = £300

Then find $\frac{1}{4}$ (which is half of half) = £150

Therefore $\frac{3}{4}$ = £450 (adding half plus a quarter)

Example 2: Finding a fraction and turning it into a percentage

There are 80 pupils in a rural junior school. 10 pupils need additional help with reading.

What is the percentage of pupils that need reading help?

The fraction of pupils that need help = $\frac{10}{80}$, by dividing top and bottom numbers by 10 we get $\frac{1}{8}$

To convert 1/8 into a percentage simply multiply 1/8 by 100

= 1/8X100 =100/8 = 50/4 =25/2 =12.5%

Money

Compound Interest, Appreciation and Depreciation

Example 1: Work out the final amount at the end of one year if there is a 10% increase (appreciation) per annum and I have £3000 to start with.

The traditional method is to work out 10% of £3000 first. Then add this answer to £3000 to get the final answer. 10% of £3000 = £300. So the final amount after a 10% increase is £3000 + £300 =£3300

Here is fast and efficient method to work out the final answer.

Simply work out 1.1X3000

Since 1.1 denotes a 10% increase.

Why 1.1? Since 100% plus 10% =1 + 0.1 = 1.1

Now 1.1× 3000 = £3300 which is the final answer

Now consider a problem involving recurring percentage changes

Example 2: Compound Interest

Find the value of £5000 if I gain a profit of 10% the first year followed by (10% of the new amount) in the second year.

THIS MEANS THE INCREASE IS 1.1X FOLLOWED BY 1.1X AGAIN

Or $(1.1)^2 \times 5000$

$(1.1)^2 \times 5000 = 1.21 \times 5000 = £6050$

So the final value is £6050

Example 3:

I buy a one bedroom apartment for £200,000. It increases in value by 5% per annum. How much will it be worth in 15 years? (**Hint: This is similar to compound interest**)

Method: Increase after 1 year will be 1.05×£200,000, after two years it will be: $(1.05)^2$ ×£200,000, after three years it will be $(1.05)^3$ ×£200,000. So, after 15 years it will be worth $(1.05)^{15}$ ×£200,000 =£415786

Example 4:

A car depreciates by 30% per annum. I buy it initially at £18000. What is its value in 5 years' time? Give your answer to the nearest pound

Method:

After one year its value will decrease by 30%, so its new value will be 70% of original as shown below:

£18000×0.7, hence, after five years its value will be £18000×$(0.7)^5$

Value after five years is £3025

Simplifying and manipulating fractions

Reducing a fraction to its lowest terms

Basically you need to find numbers that divide into the top number (numerator) as well as the bottom number (denominator), and then divide them both by the same number (start with 2, if doesn't go then choose 3, then 5, and then the next prime factor e.g. 7, 11, etc.)

Example1: Reduce $\frac{16}{24}$ to its lowest terms.

8 divides exactly into 16 and 24, so in the fraction $\frac{16}{24}$ divide top and bottom by 8. This gives the answer as $\frac{2}{3}$

In case you can't see this straight away, try starting with the number two and work your way numerically upwards using the next prime factor i.e. try 3, then 5 etc. if required

So for the fraction $\frac{16}{24}$ we can start dividing top and bottom by 2 to give us $\frac{8}{12}$, then do the same again as both 8 and 12 are still divisible by 2. This gives us $\frac{4}{6}$ and finally repeating the process once more reduces the fraction to $\frac{2}{3}$ which is the simplest form.

Example 2: Simplify $\frac{9}{12}$ to its lowest terms. In this case we can't divide top and bottom by 2, so we try 3. Since 3 will go into both 9 and 12, we can reduce this to the fraction $\frac{3}{4}$ (since 9 ÷3 =3 and 12 ÷ 3 =4) Hence, $\frac{9}{12}$ reduces to $\frac{3}{4}$

Example 3:

Reduce fraction $\frac{49}{77}$ to its lowest terms. This time we need to spot that 2, 3, 5, does not go into either 49, or 77. Either by trial and error or by spotting the right number we notice 7 goes into both the numerator and the denominator. This reduces

$$\frac{49}{77} \text{ to } \frac{7}{11}$$

> **Cancelling down fractions to their simplest form (lowest terms)**
>
> To simplify a fraction to its lowest terms you divide the numerator and the denominator by the same prime factors (2, 3, 5, 7, 11, etc.) to give the equivalent fractions as shown in the examples above

Finding fraction of an amount

Example: Find $\frac{2}{5}$ of 25

Simply replace the 'of' by ×. (times)

So $\frac{2}{5}$ of 25 becomes $\frac{2}{5} \times 25$

To work this out find out 1/5 of 25 and then multiply the answer by 2.

So 25 divided by 5, equals 5, then 2 × 5 = 10

Hence $\frac{2}{5}$ of 25 = 10

Adding and Subtracting Fractions

This next section will help you revise adding, subtracting, multiplying and dividing fractions together.

Consider adding and subtracting fractions together.

When the bottom numbers (denominators) are the same, just add the top numbers together keeping the bottom number the same. Likewise for subtraction just subtract the top two numbers.

Example 1: $\frac{2}{5} + \frac{1}{5} = \frac{3}{5}$

Example 2: $\dfrac{2}{5} - \dfrac{1}{5} = \dfrac{1}{5}$

When the denominators are different

Example 3: Work out $\dfrac{1}{2} + \dfrac{2}{5}$

When the denominators are different, the traditional method of doing this is to find the lowest common denominator. We have to find a number that both 2 and 5 will go into. This is clearly 10.

We can now re-write the fraction with the same common denominator.

To do this we have to ask how did we get the denominator from 2 to 10 for the first part, and likewise for the second part from 5 to 10. The answer is shown below:

$$\dfrac{1 X 5}{2 X 5} + \dfrac{2 X 2}{5 X 2} = \dfrac{5}{10} + \dfrac{4}{10} = \dfrac{9}{10}$$

We had to multiply top and bottom by 5 for the first part and top and bottom by 2 for the second part as shown above. We can then add the fraction as we have the same common denominator.

We can however use another very simple strategy that always works. The method is that of crosswise multiplication.

The basic method is to take the fraction sum and do crosswise multiplication as shown by the arrows. In addition, multiply the denominators (bottom numbers) together to get the new denominator.

Example 1: $\dfrac{1}{2} + \dfrac{2}{5} = \dfrac{1}{2} \times \dfrac{2}{5} = \dfrac{1 X 5 + 2 X 2}{2 X 5} = \dfrac{5+4}{10} = \dfrac{9}{10}$

We notice that if we cross multiply as shown we get 1 X 5 and 2 X 2 respectively at the top. To get the bottom number we simply multiply the bottom numbers, 2 and 5 together. So the denominator is 2 X 5=10.

Let us try another example:

Example2: Work out $\dfrac{3}{7} + \dfrac{2}{5}$

Using crosswise multiplication and adding rule, as well as multiplying the bottom two numbers we get:

$$\dfrac{3}{7} \times \dfrac{2}{5} = \dfrac{3X5+7X2}{35} = \dfrac{15+14}{35} = \dfrac{29}{35}$$

This is a very elegant method which always works

Example3: Work out $\dfrac{3}{7} - \dfrac{2}{5}$

This is similar to the above except instead of adding we now subtract as shown below.

$$\dfrac{3}{7} \times \dfrac{2}{5} = \dfrac{3X5-7X2}{35} = \dfrac{15-14}{35} = \dfrac{1}{35}$$

Note: In fact you can use this method when adding or subtracting any fraction that you find difficult. Even if you use this method for simple cases, you will still get the right answer but you may have to cancel down to get the lowest terms for the final answer.

For example we know that $\dfrac{1}{4} + \dfrac{1}{2} = \dfrac{3}{4}$

But if we didn't know and used the method shown we would get $\dfrac{1}{4} \times \dfrac{1}{2} =$
$\dfrac{1X2+4X1}{4X2} = \dfrac{2+4}{8} = \dfrac{6}{8} = \dfrac{3}{4}$ (we get this by dividing both the numerator and denominator in $\dfrac{6}{8}$ by 2). So we get the same answer in the end

Questions involving fractions

(1) Find $2\dfrac{3}{4}$ of 64

We first work out 2 X 64 = 128

To work out three quarters of 64, we first work out a half and then add it to a quarter of 64.

Half of 64 is 32

A quarter of 64 (is half of 32) is 16

Hence three quarters of 64 = 32 + 16 = 48

So two and three quarters of 64 = 128 + 48 = 176

Adding and subtracting mixed numbers

This is a similar process. We first add or subtract the whole numbers and then the fractional parts.

Example 1: $2\frac{2}{5} + 4\frac{3}{7}$

Adding the whole numbers we get 6. (Simply add 2 and 4)

Now add the fractional parts to get: $\frac{14+15}{35} = \frac{29}{35}$

So the answer is $6\frac{29}{35}$

Example 2: $4\frac{3}{7} - 2\frac{2}{5}$

Subtract the whole numbers and then the fractional parts, which gives us:

$2\frac{15-14}{35} = 2\frac{1}{35}$

Multiplying Fractions

Multiplying fractions by the traditional method is quite efficient so we will consider only this approach.

Example: $\dfrac{2}{3} \times \dfrac{5}{7} = \dfrac{10}{21}$

In this case we simply multiply the top two numbers to get the new numerator and multiply the bottom two numbers together to get the new denominator, as shown above.

Division of Fractions

When dividing fractions we invert the second fraction and multiply as shown.

Think of an obvious example. If we have to divide ½ by ¼ we intuitively know that the answer is 2. The reason for this is that there are 2 quarters in one half. Let us see how this works in practice.

Example 1: $\dfrac{6}{11} \div \dfrac{5}{11} = \dfrac{6}{11} \times \dfrac{11}{5} = \dfrac{66}{55} = \dfrac{6}{5} = 1\dfrac{1}{5}$

Step1: Re-write the fraction inverting the second fraction as shown

Step2: Multiply the top part and the bottom part to get $\dfrac{66}{55}$ **as shown.**

Step 3: Simplify this by dividing top and bottom by 11 to get $\dfrac{6}{5}$. Now this finally simplifies to $1\dfrac{1}{5}$ as shown.

The following steps are required to convert a mixed number into a fraction.

Step 1: Multiply the denominator of the fractional part by the whole number and add the numerator. Consider the mixed number $2\frac{1}{4}$. This works out to 2 × 4 + 1 = 9. This now becomes the new numerator.

Step 2: The new denominator stays the same as before. Now re-write the new fraction as $\frac{9}{4}$. (That is the new numerator ÷ existing denominator)

Example 3: Convert the mixed number, $3\frac{3}{7}$ into a fraction.

Step 1: Multiply denominator of fractional part by whole number and add the numerator. This gives 3 × 7 + 3 = 24 as the new numerator. **Step 2:** Re-write fraction as new fraction. This is now the new numerator ÷ existing denominator. This gives us $\frac{24}{7}$

Multiplying mixed numbers together

Consider the examples below:

Example: $1\frac{1}{5} \times 1\frac{3}{8}$

The method is simply to convert both mixed numbers into fractions and multiply as shown below:

$$1\frac{1}{5} \times 1\frac{3}{8} = \frac{6}{5} \times \frac{11}{8} = \frac{66}{40} = 1\frac{26}{40} = 1\frac{13}{20}$$

Notice $\frac{26}{40}$ simplifies to $\frac{13}{20}$

Dividing mixed numbers together

Example: $1\frac{1}{2} \div 1\frac{1}{4}$

There are two steps required to work out the division of mixed numbers.

Step1: Convert both mixed numbers into fractions as before

Step 2: Multiply the fractions together but invert the second one.

$$1\frac{1}{2} \div 1\frac{1}{4} = \frac{3}{2} \div \frac{5}{4} = \frac{3}{2} \times \frac{4}{5} = \frac{12}{10} = 1\frac{2}{10} = 1\frac{1}{5}$$

Proportions and ratio

Although proportion and ratio are related they are not the same thing – see example below for clarification.

Example: In a class there are 15 girls and 10 boys. The **ratio of girls to boys is** 15:10, or 3:2, (divide both 15 and 10 by 5) and the **proportion of girls in the class** is 15 out of 25 or $\frac{15}{25}$ **which simplifies to** $\frac{3}{5}$

Questions based on proportions and ratios

Example 1:

In a class of 27 pupils, 9 go home for lunch. What is the proportion of pupils in this class that have lunch at school?

Since 9 out of 27 pupils go home, this means 18 pupils have lunch at school.

As a proportion this is 18 out of 27 or $\frac{18}{27}$ which simplifies to $\frac{2}{3}$

Example 2: The ratio of boys to girls in a class is 2:3. There are 25 pupils altogether. How many boys are there?

Step 1: Find out the total number of parts, you can do this by adding up the ratio parts together. E.g. 2:3 means there are (2+3) = 5 parts altogether. This means 1 part = one fifth of 25 pupils = 5 pupils.

Since the ratio of boys to girls is 2:3, there are 2×5 boys and 3×5 girls

The number of boys in the class =2×5 = 10

Scales and ratios

Consider that you are reading a map and the scale ratio is 1: 100000

This means for every one cm on the map the actual distance is 100000 cm or put another way every one cm on the map, the distance = 1000 m (divide 100000 by 100)

to get the result in metres). Now, 1000 m = 1km (divide 1000 by 1000 to get 1 since 1km =1000m)

(Scales can also be used in other areas such as architectural drawings)

Question based on scales

I note that the map I am using has a scale of 1: 25000. The distance between the two places I am interested in is 12cm. What is the actual distance in km?

Method: 12 cm on the map corresponds to $12 \times 25000 = 300 \times 1000$ =300000cm

=3000m = 3km

(300000 /100 to convert to metres = 3000 m, now divide 3000 by 1000 to convert to km)

Hence the distance between the two places is 3km

Conversions

Conversions are often useful in changing currencies for example from pounds to dollars or euros or vice-versa. It is also useful to convert distances from miles to kilometers or weights from kilograms to pounds and so on.

Basically a conversion involves changing information from one unit of measurement to another. Consider some examples below:

Question based on conversions

Example 1: I go to France with £150 and convert this into Euros at 1.2 Euros to a pound. **(1)** How many Euros do I get? **(2)** I am left with 39 Euros when I get back home. The exchange rate remains the same. How many pounds do I get back?

Method:

(1) Since 1 pound = 1.2 Euros, I get $150 \times 1.2 = 180$ Euros in total.

2) When I get back I change 39 Euros back into pounds. This time I need to divide 39 by 1.2 So $39 \div 1.2 = 32.5$. This means I get back £32.50

Example 2:

The formula for changing kilometers to miles is given by, $M = \frac{5}{8} \times K$

Use this formula to convert 68 kilometers to miles

Method: substitute **K** with 68 and multiply by $\frac{5}{8}$

This means $M = \frac{5}{8} \times 68$. Using a calculator this comes to 42.5 miles

It is worth reviewing some common Metric and Imperial Measures as shown below

Metric Measures

1000 Millilitres (ml) =1 Litre(l)

100 Centilitres (cl) =1 Litre (l)

10ml =1 cl

1 Centimetre (cm) =10 Millimetres (mm)

1 Metre (m) = 100 cm

1 Kilometre (km) =1000 m

1 Kilogram (kg) =1000 grams (g)

1 tonne = 1000 Kg

Conversions from Metric to Imperial Measures

(You can use these values)

1 km = 5/8 mile

1 mile = 8/5 km

1kg = 2.2 pounds

1 gallon = 4.5 litres

1 inch = 2.54 cm

Sets and Venn Diagrams

Sets: A <u>set</u> is simply a collection of things and is normally represented by a capital letter. Each name, number, colour, age in the set is referred to as an **element** or member of that set. You can also represent sets by **circles in a Venn diagram.**

Example 1: A = {2, 3, 4, 5). This means that the elements (numbers in this case) are 2, 3, 4, and 5.

We can also represent this set as a circle as shown below:

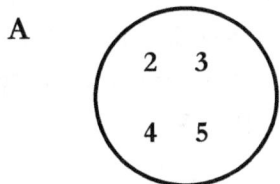

Example 2: The pupils who got grade A* in English in 2012 in a certain school were Jill, Fatima and John. We can write this as B = {Jill, Fatima, John}.

The corresponding circle would be as shown below:

Typically, you will have two or three sets that are related in some way. The rectangle that encloses all the circles is called a Universal set and is typically denoted by **E** or **ξ** or **U**. **In the exam the symbol ξ is used for a Universal Set, just remember that a <u>rectangle</u> enclosing the circle(s) represents the <u>Universal</u> Set as shown in the example on the next page.**

Example: The universal set ξ = {5, 6, 7, 8, 9, 10}. A = {5, 7} and B = {6, 7, 8, 10}. This can be represented by a Venn diagram as shown below.

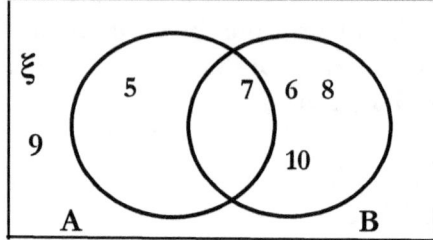

Explanation: You can see that the number **7** belongs to both set A and B (the overlapping or intersecting part of the circles). Also the number **9** is neither in set A nor in set B so it is outside both the circles but still part of the rectangle which is the Universal set ξ The remaining numbers are shown in the respective sets.

Symbols and notation associated with Sets:

You need to be familiar with some basic symbols associated with sets. For example 'U' stands for Union, so A∪B means it belongs either to set

A or B when two sets are involved. Or in the case of three sets i.e. A∪B∪C it belongs to either set A or Set B or Set C.

The symbol ∩ stands for Intersection, so that A∩B means it must be in both sets A and B (this is the overlapping part of set A and set B), in the case of two sets and in all three sets A and B and C when three sets are involved. This will be clearer in the examples shown later which illustrate these points.

Also **A ⊂ B** means A is a **proper subset** of B. This means set A is part of set B. Also note that A' means the complement of A that is all the elements that do not belong to set A.

As mentioned before, E or ξ or U refers to the universal set. This is all the elements in the rectangle. (In the exam the Universal set will be probably be referred to by the symbol ξ, **but for now do not worry about the symbol used for the Universal Set))**

∅ This symbol denotes the empty set - a set with no items or elements in it

Element of a set

∈ means it is an **element** of a set and ∉ means it is **not an element** of a set

Also n(∈) means the number of elements in a given set

Basics of Venn Diagrams

As we saw earlier a **Venn diagram is a pictorial way to represent sets.. Let us consider some examples involving just two sets.**

Example 1: A∪B (**A union B**) is simply all of A or all of B as represented by the shaded diagram below. (**Note in some of the diagrams below for simplicity I have used the notation 'U' to represent the universal set- remember the rectangle represents the Universal set, as stated before for now do not worry about the symbol used for a Universal set for now**)

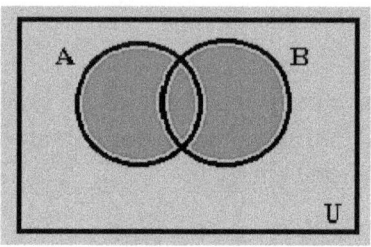

Example 2:

A ∩ B (**A intersection B**) is the dark shaded region, that is where the two circles intersect.

A ∩ B

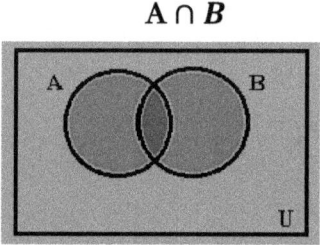

Example 3: A' or the complement of A that is everything in the Universal set but **not in A** as shown by the shaded region below:

A'

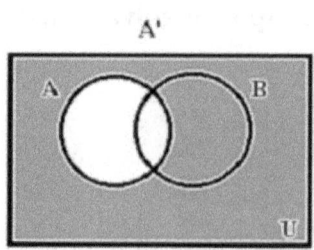

Example 4: A'∩B

The first diagram is A' (everything not in A) and the second diagram simply shows Set B. So we now have to ask where they intersect. Clearly this is the portion shown in the third diagram below.

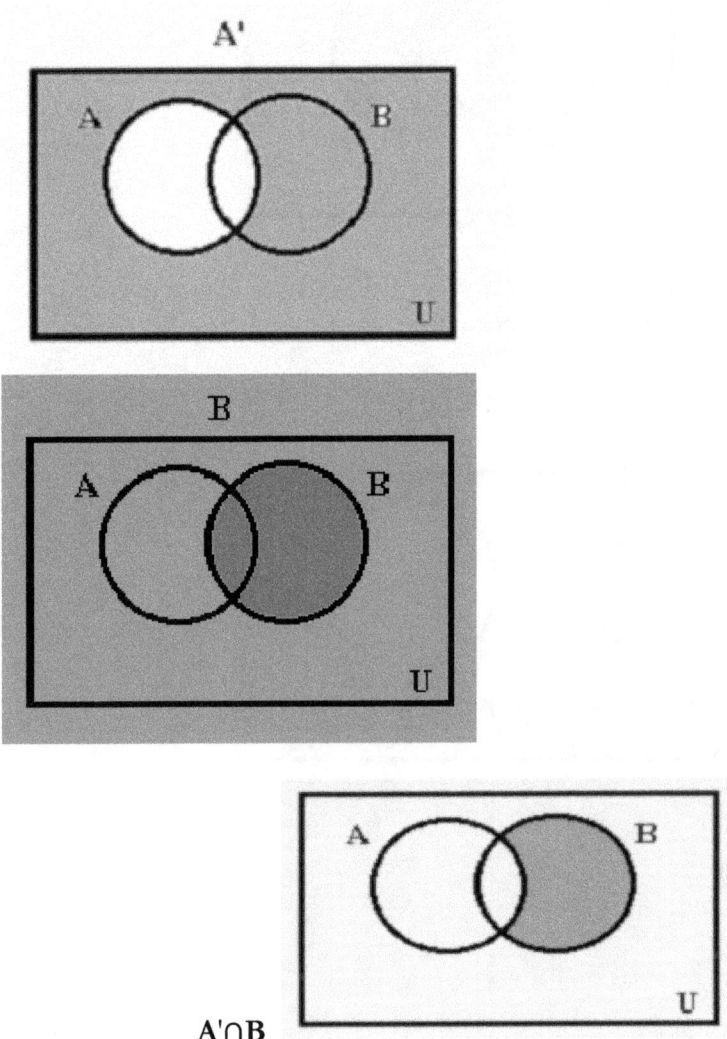

A'∩B

Example 5: (A∪B)'

First shade what's in the brackets, namely A∪B (A union B)

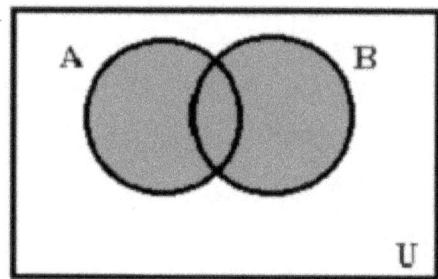

Now take the complement of the union of these sets. This is shown by the shaded bit in the diagram below:

(A∪B)'

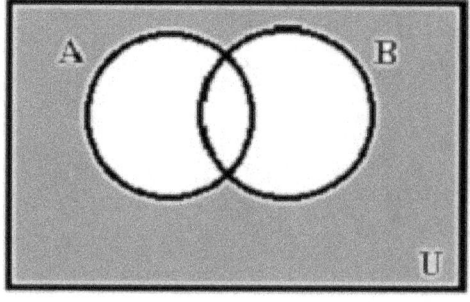

Venn Diagrams for 3 Sets

Unions, intersections and complements, are handled in the same way as they are with 2 set Venn Diagrams. For example the **intersection** of A∩B∩C is shown by the arrow where the three circles overlap.

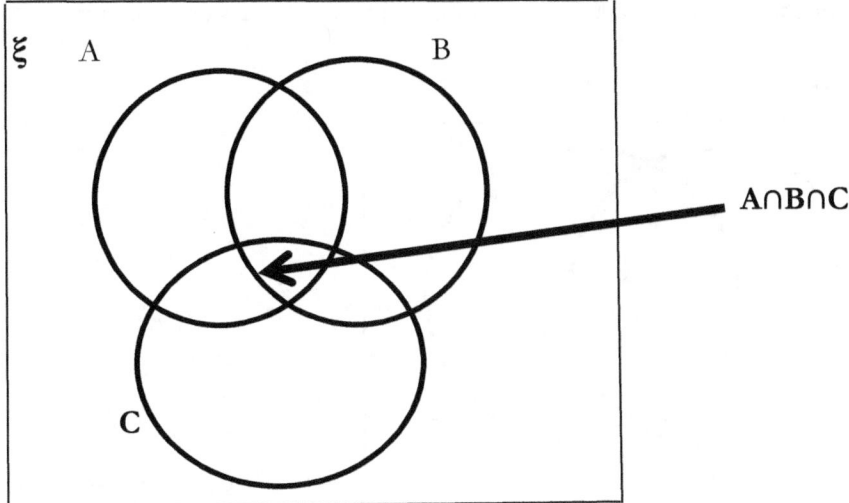

Now let us consider typical questions you are likely to get in the exam

Example 1: ξ is the universal set 1, 4, 9, 12 and 16. **A** represents the set of odd numbers and B represents the set of square numbers, show how the numbers 1, 4, 9, 12 and 16 can be illustrated by a Venn diagram.

Answer:

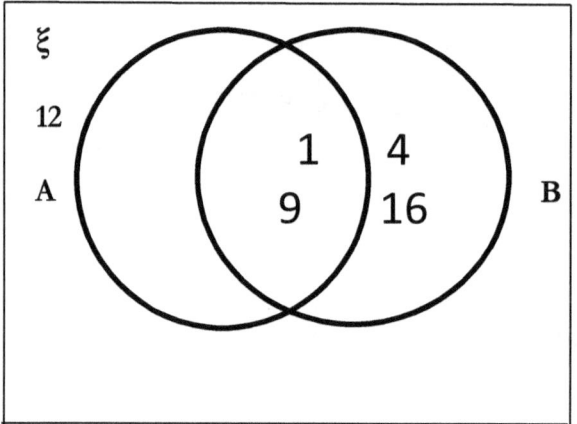

Explanation: The rectangle represents the Universal set. Circle A represents the set of odd numbers and circle B represents the set of square numbers. The numbers 1, 4, 9, 12 and 16 have to be positioned inside the Venn diagram such that the stated conditions are satisfied. Namely, 1, 4, 9 and 16 are square numbers ($1\times1 =1$, $2\times2 =4$, $3\times3 =9$ and $4\times4 =16$). 1 and 9 are odd numbers and 12 is neither odd nor square. So 4 and 16 are located inside B, 1 and 9 are located inside the intersection of A and B, and 12 is located outside both A and B.

Example 2: The circles in the Venn diagram below represent passes with GCSE grade A* in Science, Maths and English. Janet (J) passes at this grade in all these three subjects. Alex (A) passes at this grade in English and Maths but fails to get this grade in Science. Represent this information in this Venn diagram.

Answer:

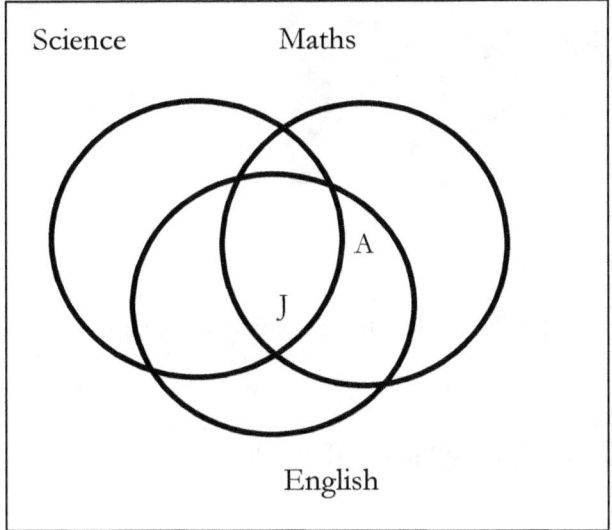

Explanation: The Venn diagram above represents Janet's (J) and Alex's (A) positions as shown.

Practice Questions 1 - <u>No Calculators allowed</u>

(1) Calculate the value of 13×120 and write the answer in standard form (2 marks)

(2) Write 220 as a product of prime factors (2 marks)

(3) Work out $3\frac{1}{3} - 2\frac{2}{7}$ (2 marks)

(4) Find the cubic root of 729 (2 marks)

(5) Find the highest common factor of 15 and 45 (2 marks)

(6) Simplify $y^{-3} \times y^5 \times \frac{1}{y}$ (3 marks)

(7) Rationalise (simplify) the surd $\frac{1+\sqrt{2}}{1-\sqrt{2}}$ (3 marks)

(8) Estimate the value of $\sqrt{15.9} \times 1.98^2$ (2 marks)

(9) Convert the recurring decimal $3\dot{2}$ to a fraction (4 marks)

(10) Using upper and lower bounds for the sides of a rectangle find the maximum area whose sides are 4cm by 2 cm. (3 marks)

(11) I buy a coat for £160 and sell it for £200. What is my percentage profit? (2 marks) (2 marks)

(12) Work out $3\frac{1}{4} \div 4\frac{3}{4}$ (3 marks)

(13) I buy a mini-ipad at a discount of 20% for £200. What was its original price? (3 marks)

(14) Given that 8km is approximately 5miles. How many kilometres are there in 75 miles? (2 marks)

(15) How many milligrams are there in 2.5 kilograms? Give your answer in standard form (3 marks)

(16) Work out $1\frac{1}{6} \times 1\frac{1}{3}$ (3 marks)

(17) Three friends go out for lunch. They buy 3 hamburgers for £1.95 each, and 3 teas for 1.15 each. What is the total cost of the meal? (2 marks)

(18) I change £52 into dollars. The exchange rate is $1.60 per pound sterling. How many dollars do I get? (2 marks)

(19) In a class of 28 pupils, 8 pupils have extra maths tuition. What is the proportion of pupils in this class that do not have extra maths tuition? (2 marks)

(20) If A = {2, 4, 7} and B = {2, 5, 7, 10}. What is A∩B? (1 mark)

Answers to Non-Calculator Practice Questions

(1) Answer: 1.56×10^3

(2) Answer: $2^2 \times 5 \times 11$

(3) Answer: $1\frac{1}{21}$

(4) Answer: 9

(5) Answer: 15

(6) Answer: y

(7) Answer: $-3 - 2\sqrt{2}$

(8) Answer: Approximately 16

(9) Answer: $\frac{32}{99}$

(10) Answer: $11.25 cm^2$

(11) Answer: 25%

(12) Answer: $\frac{13}{19}$

(13) Answer: £250

(14) Answer: 120km

(15) Answer: 2.5×10^6 mg

(16) Answer: $1\frac{5}{9}$

(17) Answer: £9.30

(18) Answer: $83.20

(19) Answer: $\frac{5}{7}$

(20) Answer: A∩B = {2, 7}

Practice Questions 2: Calculators allowed

(1) £280 is divided in the ratio 5: 2: 1. Find the largest part. (2 marks)

(2) I put £1500 in a bank deposit account which gives me a return of 3% per annum compound. How much in total will I have in 3 years? (3 marks)

(3) Work out $3\frac{1}{3} - 2\frac{2}{9}$ (3 marks)

(4) Simplify $(y^{-3} \times y^5) \div y^7$ (3 marks)

(5) I travel 500km on a holiday. How much is this in miles? (2 marks)

(6) I convert 150 euros to pounds when I come back from a holiday in France. The exchange rate is 1.28 euros to a pound. How much do I get back in pounds and pence? (3 marks)

(7) A map that I am using has a scale of 1: 50000. The distance between the two places I am interested in is 8.5cm. What is the actual distance in km? (3 marks)

(8) In a concert hall which has a seating capacity for 1300 persons, 350 seats are vacant during a particular concert. What is the proportion of seats that is vacant? Give your answer as a fraction in its lowest terms. (3 marks)

(9) Work out $(1\frac{1}{4} \times 1\frac{3}{8}) \div 2\frac{3}{16}$ (4 marks)

Answers to Calculator Practice Questions

(1) Answer: The largest part is £175

(2) Answer: £1639.10

(3) Answer: $1\frac{1}{9}$

(4) Answer: y^{-5} or $\frac{1}{y^5}$

(5) Answer: 312.5 miles

(6) Answer: £117.19

(7) Answer: 4.25 Km

(8) Answer: $\frac{7}{26}$

(9) Answer: $\frac{11}{14}$

Basic Algebra

In algebra we often use letters instead of numbers. There are some basic conventions and rules of algebra that you should be familiar with to progress in this subject. This chapter will be useful for you if you have forgotten your algebra.

If you see	we mean
$x = y$	x equals y
$x > y$	x is greater than y
$x < y$	x is less than y
$x \geq y$	x is greater than or equal to y
$x \leq y$	x is less than or equal to y
$x + y$	the sum of x and y
$x - y$	subtract y from x
xy	x times y
x/y	x divided by y
$x \div y$	x divided by y
x^n	x to the power n
$x(x + y)$	x times the sum of $x + y$

Also note that:

$x(x + y) = x^2 + xy$

$$x^2(x + x^2 + y) = x^3 + x^4 + x^2y$$

In general, a x a x a x a ....(n times) $= a^n$

You also need to know these algebraic rules for the multiplication and division of positive and negative numbers.

Multiplying positive and negative numbers.

$(+) \times (+) = +$ (a plus number times a plus number gives us a plus number)

$(+) \times (-) = -$ (a plus number times a minus number gives us a minus number)

$(-) \times (+) = -$ (a minus number times a plus number gives us a minus number)

$(-) \times (-) = +$ (a minus number times a minus number gives us a plus number)

Dividing positive and negative numbers.

$(+) \div (+) = +$ (a plus number divided by a plus number gives us a plus number)

$(+) \div (-) = -$ (a plus number divided by a minus number gives us a minus number)

$(-) \div (+) = -$ (a minus number divided by a plus number gives us a minus number)

$(-) \div (-) = +$ (a minus number divided by a minus number gives us a plus number)

Summary: <u>For both multiplication and division, like signs gives us a plus sign and unlike signs gives a minus sign</u>

Also when adding and subtracting it is worth knowing that:

When you add two minus numbers you get a bigger minus number.

Example 1: $-4 - 6 = -10$

When you add a plus number and a minus number you get the sign corresponding to the bigger number as shown below:

Example 2: $+6 - 9 = -3$, whereas, $-6+9 = 3$

When you subtract a minus from a plus or minus number you need to note the results as shown below:

Example 3: $6 - (-3)$ we get $6+3 = 9$ (since $-(-3) = +3$)

Example 4: $7 - (+3)$ we get $7 - 3 = 4$ (since $-(+3) = -3$)

In this case note that $-(-) = +$. Also, $+(-) = -$ and $-(+) = -$.

Simplifying algebraic expressions

Example 1: Simplify $3x + 4x + 5x$

Method: We simple add up all the x's.

Hence we get $3x+4x+5x = 12x$

Example 2: Simplify $3x + 4x + 3y + 5y$

Method: Add up all the like terms.

So we get $3x+4x +3y+5y = 7x +8y$

(Notice we add up all the x's and then all the y's)

Example3: Simplify $3m + 4y + 2m - 3y$

Method: as before, we add and subtract like terms.

Now $3m+2m = 5m$ and $4y-3y = 1y$ or just y.

So we can write $3m + 4y + 2m - 3y = 5m + y$.

Multiplying out brackets.

Example 1: Expand and simplify $3(2x+5) + 4(2x+7)$

Method: Multiply 3 by each term in the first bracket then 4 by each term in the second bracket. The final step is to simplify by collecting up the like terms.

$3(2x+5) + 4(2x+7) = 6x + 15 + 8x + 28 = 14x + 43$

Notice the last step is simply adding $6x + 8x$ and then $15+28$.

Example 2: Work out $(2x+3)(2x+4)$

When we have to multiply out two brackets we have to multiply each term in the first bracket by each term in the second bracket. We then simplify the resulting expression as before. An easy way to multiply out two brackets is to use the grid method as shown below:

First put each of the terms of each bracket on the outside grid as shown

X	2x	+3
2x		
+4		

Step2: Multiply each outside term together. So that for example $2x \times 2x = 4x^2$. The other results are shown inside the grid.

X	2x	+ 3
2x	$4x^2$	+ 6x
+ 4	8x	+12

After multiplying out the terms, the answer is found by adding all the terms inside the grid and simplifying the resulting expression.

So we have, $4x^2 + 6x + 8x + 12$ (These are all the terms inside the grid)

Finally, $4x^2 + 6x + 8x + 12 = 4x^2 + 14x + 12$

Another way of expanding brackets

Example 1: Expand $(x + 3)(x + 2)$

(Multiply the first term of the first bracket by the second bracket and then multiply the second term of the first bracket by the second bracket. Finally simplify the expression.)

So $(x + 3)(x + 2) = x(x + 2) + 3(x + 2) = x^2 + 2x + 3x + 6 = x^2 + 5x + 6$

Example 2: Expand $(2x - 1)(x - 2)$

This equals $2x(x - 2) - 1(x - 2) = 2x^2 - 4x - x + 2 = 2x^2 - 5x + 2$

Simplifying Algebraic Fractions

Example 1: Simplify $\frac{1}{x+3} + \frac{2}{3}$

Method: First find the common denominator which is x(x + 3)

Then treat it like you were simplifying a fraction

$$\frac{1}{x+3} + \frac{2}{3} = \frac{1\times 3 + 2(x+3)}{3(x+3)} = \frac{3+2x+6}{3(x+3)} = \frac{9+2x}{3(x+3)}$$

Example 2: Simplify $\frac{2}{x-3} - \frac{1}{5}$

As before $\frac{2}{x-3} - \frac{1}{5} = \frac{5\times 2 - 1(x-3)}{5(x-3)} = \frac{10-x+3}{5(x-3)} = \frac{13-x}{5(x-3)}$

Factorising

Example 1: Factorise: $3x^2 - 6xy$

= 3x(x − 2y) (find the common factor which is 3x in this case)

Example 2: Factorise: $3t^2 y - 9t^3$

= $3t^2$(y − 3t)

Something useful to remember is the <u>difference of two squares:</u>

$p^2 - q^2 = (p + q)(p - q)$

Since (p + q)(p − q) = p(p − q) + q(p − q) = p^2 + pq − pq - q^2 = = p^2 - q^2

So for example: $16x^2 - 9y^2$ = (4x − 3y)(4x + 3y)

Algebraic Substitution and Formula

This is the process of substituting numbers for letters and working out value of the corresponding expression. We have already met substitution in the section on Formulas. But here are more examples that will clarify the process.

Example 1: If k=6 and t=8 work out 2(4k–2t) +kt

Substituting the values of k and t we have:

2(4 × 6–2 × 8) + 6 × 8

=2× (24 – 16) +48 = 2× 8 +48 =16+48 =64

So 2(4k – 2t) + kt = 64

Example 2: If t=9 and u= 6 work out $3t^2$ -5u

Substituting appropriately we get:

3 X 9^2 - 5 X 6 = 3 X 81–30 =243-30 =213

(Notice, we use the BIDMAS rule to work out the square first and then do the multiplication)

So, $3t^2$ -5u =213

Formula

A formula describes the relationship between two or more variables. You have seen some examples above already. But now let us consider some practical examples.

Example 1:

(1) The formula for working out the distance depends on the speed and time taken in the appropriate units.

D = S×T where D is the distance, S the speed and T is the time.

What is the distance travelled if my speed is 60kmh and I travel for 1hour and 30 minutes.

1 hour 30 minutes corresponds to 1.5 hours so, using the formula, D = 60×1.5 = 90 km

That is, the distance equals 90km

(2) The formula for working out the speed is given as Speed= Distance/Time

That is S = D÷T

Work out the average speed with which I travel, if I cover 100 miles in 2.5 hours.

Since S = D÷T, this means S = 100÷2.5 =40 mph (Notice the units for the first example were in kilometres and units for the second example were in miles)

Example 2:

The formula for converting the temperature from Celsius to Fahrenheit is given by the formula: $F = \frac{9}{5}C + 32$ (where C is the temperature in degrees Centigrade)

If the temperature is 10 degrees Celsius then what is the equivalent temperature in Fahrenheit?

Using the formula $F = \frac{9}{5}C + 32$, and substituting 10 in place of C, we have $F = \frac{9}{5} \times 10 + 32 = \frac{90}{5} + 32 = 18 + 32 = 50$. Hence, 10 degrees centigrade = 50 degrees Fahrenheit

Explanation of working out above: Remember we multiply and divide before adding and subtracting) There are no brackets to worry about. When working out $\frac{9}{5} \times 10 + 32$, multiply 9 by 10 to get 90, divide this by 5 to get 18, finally add 18 and 32 together to get 50

Example 3: Convert 68 degrees Fahrenheit to degrees Celsius. The formula for converting the temperature from Fahrenheit to Celsius is given by:

$C = \dfrac{5}{9}(F-32)$, So to change 68 degrees Fahrenheit to degrees Celsius we can substitute for F in the formula $C = \dfrac{5}{9}(F-32)$, $C = \dfrac{5}{9}(68-32) =$

$\dfrac{5}{9} \times 36 = 5 \times 4 = 20$. Hence, 68 degrees Fahrenheit = 20 degrees Celsius

Explanation of the working out above: Using BIDMAS we work out the bracket first. This gives us 68-32 = 36. We now divide this by 9 and multiply by 5. Clearly 36÷9 = 4 and finally 5×4 = 20

We have seen that formulas can be important in conversion problems

Earlier we saw the formula: S = D ÷T, that is, Speed $= \dfrac{\text{Distance}}{\text{Time}}$.
Sometimes in the questions you may be shown a distance time graph for a school coach trip and asked to work out average speed for a particular part of the journey and the time the coach was stationary. See example below.

Example: A school trip by coach to a heritage site leaves at 1200hrs from the school. The coach arrives at the destination at 1300hrs. It then stops so the pupils can look around the site. Finally after looking around the site it leaves and arrives back at school at 15:30hrs. (1) How long did the coach stop for? (2) What was the average speed on the return journey?

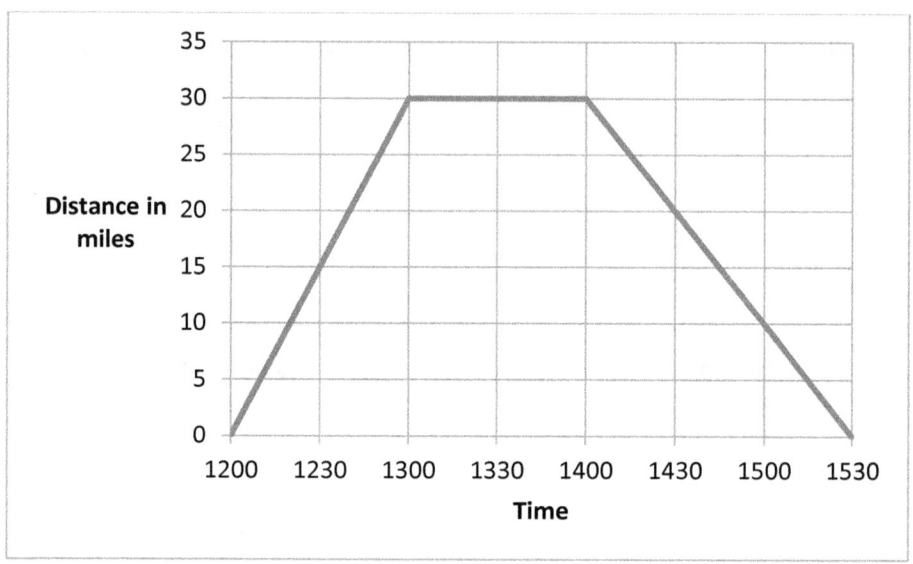

(1) From the distance-time graph above you can see it was stationary from 1300 – 1400hrs, which is 1hr

(Between these time intervals no further distance is covered, so it is stationary – see the vertical axis at 30 miles)

(2) The return journey starts at 1400hrs and ends at school at 1530hrs = 1.5 hrs.

Since ***Speed*** $= \frac{Distance}{Time}$, this means speed = 30÷1.5 = 20 mph

Solving equations

Linear Equations

Example 1: Solve the equation 2x +5=11 using an algebraic method.

Subtract 5 from both sides (that is, take the inverse of +5)

So, 2x =6

Now divide both sides by 2(that is, take the inverse of X2)

So, x =3.

Now consider slightly harder equations:

Example 1: Solve the equation 5x – 1 = 2x +8

First add 1 to both sides, which gives:

5x = 2x +9

Now subtract 2x from both sides to give 3x = 9

Finally divide both sides by 3 to get x=3.

(Notice each step simplifies the equation further)

Example 2: Solve the equation 5(2x +1) =4(2x +1)

To solve this first multiply out the bracket which gives:

10x +5 = 8x +4

(Multiply each term outside the bracket by each term inside the bracket)

Now subtract 5 from both sides, which gives:

10x =8x –1

Now subtract 8x from both sides, which gives:

2x = –1

Finally, divide both sides by 2 to get x= –1/2 or –0.5

Example 3: Solve the equation $\dfrac{2x}{3} + 5 = 7$

We can simplify this to $\dfrac{2x}{3} = 2$ (by subtracting 5 from both sides)

Now multiply both sides by 3 to get the expression below:

2x =6

So x =3

Example 4

Solve the equation $\sqrt{4 - \dfrac{x+3}{3x+2}} = 3$

Although this might look complicated the basic rule is whatever you do to one side you must do the same to the other.

Step 1: Square both sides so we get $4 - \dfrac{x+3}{3x+2} = 9$

Step 2: Cross –multiply everything by the denominator

We get: 4(3x + 2) – (x + 3) = 9(3x +2)

Simplify to get 12x + 8 – x – 3 = 27x + 18

Simplify further to get 11x + 5 = 27x + 18

Subtract 18 from both sides to get 11x – 13 = 27x

Now subtract 11x from both sides to get – 13 = 16x

Which is the same as 16x = -13, this means $x = \frac{-13}{16}$

Word problems that can be solved by algebra:

Example 1:

John has £22 more than Brian. Altogether they have £68. How much do they each have?

Let the amount Brian has be £x. Since Brian's amount plus John's amount = £68, we can write algebraically that x + x+ 22 = 68.

Simplifying this expression we get 2x + 22 =68

Subtracting 22 from both sides we get:

2x = 46, now divide both sides by 2. We get x = 23

Hence Brian has £23, and John has £45

Example 2:

A teacher buys a pay as you go smartphone that has a 60% discount. The amount she pays is £60. What was the original price of this smartphone?

Method:

Let the original price be £x

Clearly she pays 40% of original price (since the discount is 60%). This means original price is 40% of £x. We can now write this as an equation:

40% of £x = £60, that means $\frac{40}{100} \times x = 60$, simplifying this we get $\frac{4}{10} \times x = 60$

This means 4x = 600 (multiply both sides by 10 in the previous equation). Now divide both sides by 4 to get, $x = \frac{600}{4} = \frac{300}{2} = 150$. This means the original price was £150.

Simultaneous Equations

We saw earlier that simple equations allow us to solve problems involving one unknown. When you have to solve problems involving more than one unknown you need more than one equation to solve these.

A simultaneous equation with two variables (meaning two unknowns), say x and y typically involves two equations. The problem is then to find the values of x and y which satisfies the equations at the same time. Another way of saying simultaneous is 'at the same time'.

We will first consider one traditional method of solving simultaneous equations.

Example 1: Solve the simultaneous pair: $2x + 4y = 5$

$$3x + 5y = 9$$

First let us understand the problem. The problem is to find the values of x and y such that the equations are true.

Method : Eliminate one of the variables

Consider the simultaneous pair of equations as before:

$2x + 4y = 5$ (1)

$3x + 5y = 9$ (2)

Try and make the x terms or the y terms the same and then add or subtract the equations to eliminate one of the variables

Suppose we make the 'x' term the same. We multiply equation (1) by 3 and equation (2) by 2. The new equations are now shown below:

$6x + 12y = 15$ (3)

$6x + 10y = 18$ (4)

Now subtract (4) from (3) and we get:

$6x - 6x + 12y - 10y = 15 - 18$

$2y = -3$ hence $y = -3/2 = -1.5$

Now substitute $y = -1.5$ in equation (1) and we get:

$2x + 4\times(-1.5) = 5$ which means $2x - 6 = 5$ or $2x = 11$ so $x = 11/2$ or 5.5

Hence as before $x = 5.5$ and $y = -1.5$

Check

We can check in equation 1 to see if the values we found satisfy the equation.

The first equation is: $2x + 4y = 5$

Substituting, $x = 5.5$ and $y = -1.5$ we get:

$2\times5.5 + 4\times(-1.5) = 11 - 6 = 5$ as required.

Solving Word Problems using Algebra

Examples

(1) Fatima and Louise have £350 between them. Louise has £80 less than Fatima. How much do they each have?

Method: Let the amount Fatima has be represented by x
Hence, Louise has x – 80. We know that the sum of the two amounts = £350. That is x + x – 80 =350. Simplifying, we get 2x – 80 =350. Now add 80 to both sides so we have 2x - 80 +80 = 350 + 80. Which means 2x = 430, or x = 215. This means Fatima has £215 and Louise has £135 (Since Louise has £80 less than Fatima)

(2) The cost of a coat after a 20% discount is £85. What was its original price?

Method: Let the original price be £x. This means x – 20% of x = 85. Or x – 0.2x = 85, which simplifies to 0.8x =85. Now divide both sides by 0.8. So we get x = 85÷0.8 = 106.25. Hence the original price is £106.25

(3) The area of a rectangle is $162m^2$. The length of the rectangle is two times the width. What is the length and width of the rectangle?

Method: Let the width =w, hence the length = 2w. We know that the area of a rectangle is length × width = 2w×w = $2w^2$. The area of the rectangle is given as $162m^2$. Hence, $2w^2$ = 162. Dividing both sides by 2, we get w^2= 81. Hence w = $\sqrt{81}$ =9. So the width is 9m and the length is 18m.

(4) John's annual salary is $\frac{3}{4}$ of Hilary's salary. Hilary's salary is twice Betty's. The total salary between them is $450,000. How much did each of them earn?

Method: Let Hilary's salary be x (in dollars). Hence, John's salary is $\frac{3}{4}$ x. Also, since Hilary earns twice as much as Betty, then Betty earns half of

Hillary's $= \frac{1}{2}x$. Finally, we know that $x + \frac{3}{4}x + \frac{1}{2}x = \$450{,}000$, simplifying $2\frac{1}{4}x = 450{,}000$. Or, $\frac{9}{4}x = 450{,}000$. This means $9x = 1{,}800{,}000$ or $x = 200{,}000$. So Hilary earns \$200,000, John earns \$150,000 (three quarters of Hilary's amount) and Betty earns \$100, 000 (half of Hilary's salary)

Solving Quadratic Equations

The formula method of solving quadratic equations

Consider the general quadratic equation $ax^2 + bx + c = 0$

Dividing through by 'a' we get:

$$x^2 + \frac{b}{a}x + \frac{c}{a} = 0$$

Now we use the method of completing the square

First halve the middle term coefficient and then square the expression on the left hand side as shown below:

$$(x + \frac{b}{2a})^2 = x^2 + \frac{b}{a}x + \frac{b^2}{4a^2}$$

Adjusting to get the original expression, we have:

$$(x + \frac{b}{2a})^2 - \frac{b^2}{4a^2} + \frac{c}{a} = x^2 + \frac{b}{a}x + \frac{c}{a}$$

We can write $(x + \frac{b}{2a})^2 - \frac{b^2}{4a^2} + \frac{c}{a} = 0$

$$(x + \frac{b}{2a})^2 = \frac{b^2}{4a^2} - \frac{c}{a}$$

Simplifying the right hand side we get:

$$(x + \frac{b}{2a})^2 = \frac{b^2}{4a^2} - \frac{4ac}{4a^2}$$

$$(x + \frac{b}{2a})^2 = \frac{b^2 - 4ac}{4a^2}$$

$$x + \frac{b}{2a} = \pm \frac{\sqrt{b^2 - 4ac}}{2a}$$

$$x = -\frac{b}{2a} \pm \frac{\sqrt{b^2 - 4ac}}{2a}$$

$$x = \frac{-b \pm \sqrt{b^2 - 4ac}}{2a}$$

Example:

Solve the equation $2x^2 - 5x + 2 = 0$ using the quadratic formula.

Method:

When the above equation is compared to the general equation

$ax^2 + bx + c = 0$

We can see that a = 2, b = -5 and c = 2

Since, $x = \frac{-b \pm \sqrt{b^2 - 4ac}}{2a}$

By substituting the above values we can see that:

$$x = \frac{-(-5) \pm \sqrt{(-5)^2 - 4 \times 2 \times 2}}{2 \times 2}$$

$$x = \frac{5 \pm \sqrt{25 - 16}}{4}$$

$$x = \frac{5 \pm \sqrt{9}}{4}$$

$$x = \frac{5 \pm 3}{4} = \frac{8}{4} \text{ or } \frac{2}{4}$$

Hence $x = 2$ or $\frac{1}{2}$

(Note: The formula method is particularly useful if you find it hard to factorise or if a quadratic expression cannot be factorised)

Solving quadratic equations using factorisation when possible:

Example 1: Solve the equation $x^2 + 5x + 6 = 0$

We can factorise the above quadratic equation as $(x + 3)(x + 2) = 0$

This means either $x + 3 = 0 \implies x = -3$ or $x + 2 = 0 \implies x = -2$

Example 2: Solve the quadratic equation $2x^2 - 5x + 2 = 0$

We can write the above equation as $(2x - 1)(x - 2) = 0$

(You can do this by trial and error with a little bit of common sense)

For example the only way to get $2x^2$ is by having x and 2x in the two brackets. Also the only way to get + 2 as the last term is to have +1 and +2 or -1 and -2. Finally, as the middle term is -5x the factors have to be $(2x - 1)(x - 2)$

So if $(2x - 1)(x - 2) = 0$ this means either $2x - 1 = 0$ or $x - 2 = 0$

If $2x - 1 = 0 \implies 2x = 1$ and $x = \frac{1}{2}$ and if $x - 2 = 0 \implies x = 2$

Hence the solution to the quadratic equation $2x^2 - 5x + 2 = 0$ is when

Either and $x = \frac{1}{2}$ or $x = 2$

Things to note in quadratic equations and the quadratic formula:

(1) $ax^2 + bx + c = 0$ is a quadratic equation providing 'a' is not zero.

(2) There are two solutions (or roots) to a quadratic equation

(3) The roots are real so long as in the formula, $x = \dfrac{-b \pm \sqrt{b^2 - 4ac}}{2a}$ the bit inside the square root is > 0. The bit inside the square root, that is: $b^2 - 4ac$ is called the <u>discriminant</u>. Note if $b^2 - 4ac = 0$, there is only one real root

(4) When $b^2 - 4ac < 0$, then the roots are not real.

<u>Below are examples of equations with two solutions (two roots), one solution (one root) and no solution (no real roots)</u>

(a) The equation $x^2 + 2x - 15 = 0$, has two real roots, x= -5 and x = 3 as shown in the graph below

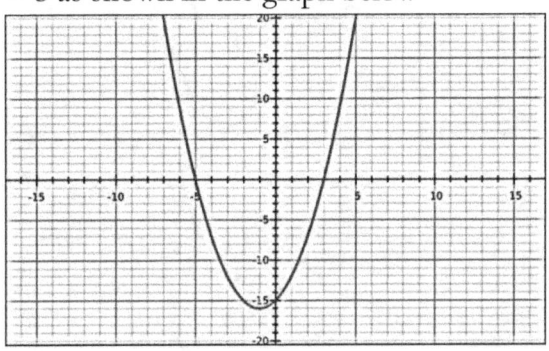

(b) The equation $x^2 - 6x + 9 = 0$ has one real root at x = 3 as shown below

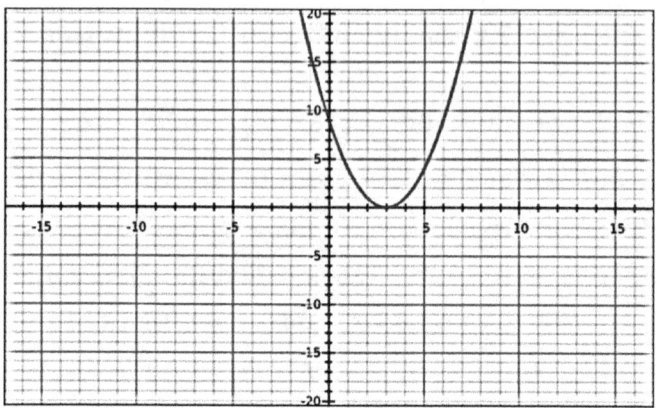

(c) The equation $x^2 - 6x + 12 = 0$ has no real roots as the parabola does not intersect the x-axis at any point.

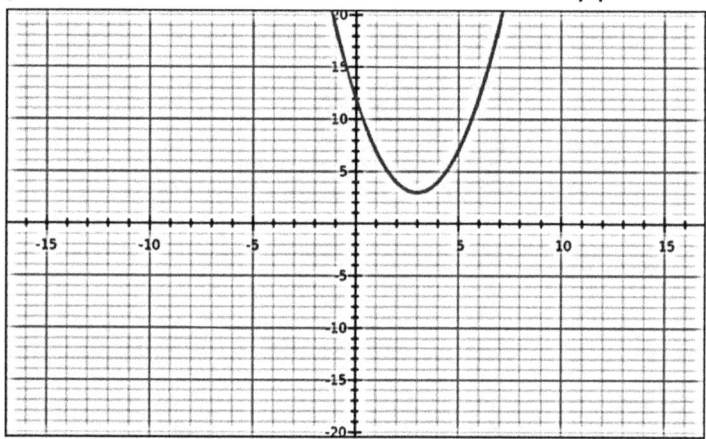

Solving Cubic equations

Using Trial and Improvement

Trial and Improvement

This is an approach to get closer and closer to the solution of an equation which might be hard to solve. The answer is always approximate but typically very close to the actual solution.

Normally for GCSE maths you are given a cubic equation to solve.

Example: Solve the equation $x^3 + 3x = 19$. You are given that the solution lies between 2 and 3. Use trial and improvement to find a solution to this equation correct to 1 decimal place.

Method: Keep trying different values of x as shown in the table below recording your answers and your comments.

x	$x^3 + 3x$		Comment
2	$2^3 + 3\times 2 = 14$		Too small
3	$3^3 + 3\times 3 = 36$		Too big
2.5	$2.5^3 + 3\times 2.5 = 23.125$		Too big
2.4	$2.4^3 + 3\times 2.4 = 21.024$		Too big
2.3	$2.3^3 + 3\times 2.3 = 19.067$		About right

Hence the solution to this equation is x = 2.3 to 1 decimal place.

Change the subject of a formula

Changing the subject of a formula (re-arranging formulas)

Example 1: In the formula $a = bx + c$ make x the subject

Method: Apply the same rules as you would to equations

In this case subtract c from both sides to get $a - c = bx$

Now divide both sides by b to get $\frac{a-c}{b} = x$

In other words $x = \frac{a-c}{b}$

Example 2: In the formula $\frac{ay^2}{b} + m = k$, make y the subject

Method:

Step 1: Subtract m from both sides to get $\frac{ay^2}{b} = k - m$

Step 2: Multiply both sides by 'b' to get $\frac{ay^2}{\cancel{b}} \times \cancel{b} = b(k - m)$

(The 'b' s on the left hand side cancel)

Hence we now have: $ay^2 = b(k - m)$

Step 3: divide both sides by 'a' (to cancel the 'a' on the left hand side)

We now have $y^2 = \frac{b(k-m)}{a}$

Step 4: Take the square root of both sides to get $y = \sqrt{\frac{b(k-m)}{a}}$

So we finally find that $y = \sqrt{\dfrac{b(k-m)}{a}}$

Example 3: In the formula $\dfrac{a}{1+t^2} = b + c$, make t the subject

Step 1: Multiply both sides by $1 + t^2$ to get: $a = (b + c)(1 + t^2)$

Step 2: Divide both sides by $(b + c)$ to get: $\dfrac{a}{b+c} = 1 + t^2$

Step 3: Subtract '1' from both sides to get: $\dfrac{a}{b+c} - 1 = t^2$

Step 4: This simplifies to $\dfrac{a-1(b+c)}{b+c} = t^2$, which simplifies to $\dfrac{a-b-c}{b+c} = = = t^2$

Step 5: Take the square root of both sides: $\sqrt{\dfrac{a-b-c}{b+c}} = t$

So finally we have $t = \sqrt{\dfrac{a-b-c}{b+c}}$

Proportionality

You need to be familiar with two basic types of proportionalities.

Directly proportional

$y \propto x$ means when x is multiplied by a number then the value of y goes up proportionately. We can re-write $y \propto x$ as $y = kx$ (k is referred to as the constant of proportionality)

Example 1: y is directly proportional to x. When x = 5, y = 20 find the value of x when y =16.

Method: Step 1: We know $y \propto x$ $\Rightarrow$ y =kx

Step 2: Find the value of k by substituting the values of x and given

$20 = k \propto 5$ or $20 = 5k$ $\Rightarrow$ Hence k = 4 (divide both sides of the previous equation by 5)

So the equation now becomes: y = 4x

Step 3: To find x when y = 9, we simply substitute the values in the above equation.

This means: $16 = 4x$. $\Rightarrow$ x = 4

Example 2: y is directly proportional to the square of x. When x = 9 then y =27. Find the value of y when x = 12.

Step 1: $y \propto x^2$ $\Rightarrow$ $y = kx^2$

Step 2: Find the value of k by substituting the initial values of x and y.

$\Rightarrow$ $27 = k9^2$ $\Rightarrow$ $27 = 81k$ $\Rightarrow$ $k = 27 \div 81$ $\Rightarrow$ $k = \frac{27}{81} = \frac{3}{9} = \frac{1}{3}$

So the equation $y = kx^2$ can be written as $y = \frac{1}{3}x^2$

Step 3: To find y, we simply substitute the value of x $\Rightarrow$ $y = \frac{1}{3} \times 12^2$
$= \frac{1}{3} \times 12^2 = \frac{1}{3} \times 12 \times 12 = 4 \times 12 = 48$

Hence when x = 12, y = 48

Inversely Proportional

$y \propto \frac{1}{x}$ this means if x increases then y decreases

$y \propto \frac{1}{x}$ $\Rightarrow$ $y = \frac{k}{x}$

Example 1: Given that y is inversely proportional to x and x = 7 when y = 3. Find the value of x when y = 0.5.

Step 1: $y \propto \frac{1}{x}$ $\Rightarrow$ $y = \frac{k}{x}$

Step 2: Substituting the initial values of x and y we get $3 = \frac{k}{7}$ $\Rightarrow$ k = 21

So the equation $y = \frac{k}{x}$ becomes $y = \frac{21}{x}$

Step 3: We can now find the value of x when y = 0.5 by substituting this value in the equation $y = \frac{21}{x}$ Hence, $0.5 = \frac{21}{x}$ $\Rightarrow$ x×0.5 = 21   x $= \frac{21}{0.5} = 42$. So when y = 0.5, x = 42

Example 2:

You are given that y is inversely proportional to $\sqrt{x}$. Also when x = 49, y = 4. Find the value of y when x = 196.

Step 1: We can write $y \propto \frac{1}{\sqrt{x}}$ $\Rightarrow$ $y = \frac{k}{\sqrt{x}}$

Step 2: Find the value of k by substituting the initial values of x and y.

$$\Rightarrow 4 = \frac{k}{\sqrt{49}} \Rightarrow 4 = \frac{k}{7} \Rightarrow k = 28$$

Step 3: We can now find the value of y when x = 196

Since $y = \frac{28}{\sqrt{x}}$, $y = \frac{28}{\sqrt{196}} \Rightarrow y = \frac{28}{16} = \frac{7}{4} = 1\frac{3}{4}$

This means when $x = 196$, $y = 1\frac{3}{4}$

Practice Questions 3 -No calculators allowed

(1) If a = 3, b = 4 and c = 5 work out 3a + 5b − 2ac (2 marks)

(2) Expand and simplify the expression 3(5x + 7) + 4(x − 8) (2 marks)

(3) Expand and simplify the expression (3x − 5)(2x + 7) (3 marks)

(4) Solve the equation: 3x − 6 = 2x + 8   (2 marks)

(5) Solve the inequality 2x + 7 > 3x + 9   (3 marks)

(6) Simplify $\frac{1}{x+6} - \frac{2}{7}$ (4 marks)

(7) By choosing suitable values of x, plot a graph of the equation
y = 2x − 8   (2 marks)

(8) Plot the graph for the equation y = $3x^2$ + 2x − 1 for values of x from
−3 to +3   (4 marks)

(9) Factorise the expression $5xy + 10x^2y^3$ (2 marks)

(10) John and Mary have £360 between them. Mary has £80 more than John. How much do they each have? (3 marks)

Answers to non-calculator practice questions:

(1) Answer: -1
(2) Answer: $19x - 11$
(3) Answer: $6x^2 + 11x - 35$
(4) Answer: $x = 14$
(5) Answer: $x < -2$
(6) Answer: $\dfrac{-5-2x}{7(x+6)}$
(7) Answer: See graph below:

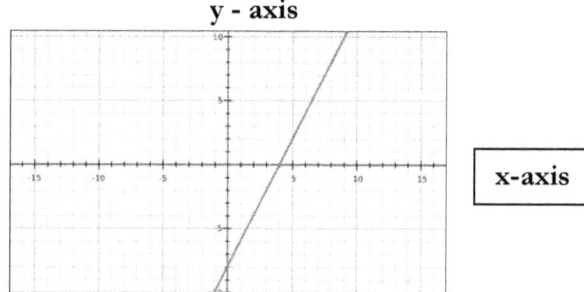

(8) Answer: See graph below:

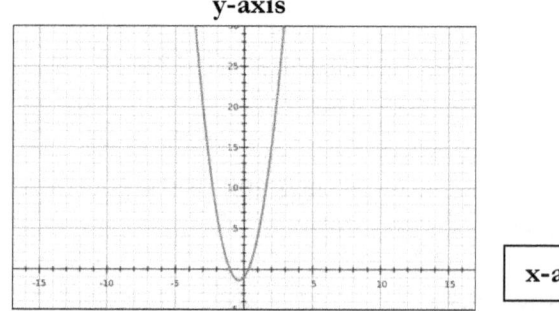

(9) Answer: $5xy(1 + 2xy^2)$
(10) Answer: Mary has £220 and John has £140

Practice Questions 4: calculators allowed

(1) You are given that m is directly proportional to $\sqrt{n}$, further when n =36, m= 3. (a) Find the constant of proportionality. (b) Find the value of m when n =25
(4 marks)

(2) Solve the quadratic equation $6x^2 + 7x - 3 = 0$ either by factorisation or using the quadratic formula   (3 marks)

(3) Solve the simultaneous pair x + 2y = 2 and 2x − 2y = 1 (3 marks)

(4) Solve the equation $x^3 + 2x = 22$. You are given that the solution lies between 2 and 3. Use trial and improvement to find a solution to this equation correct to 2 decimal places. (3 marks)

(5) A pupil buys an app for her phone that has a 40% discount. The amount she pays is £3. What was the original price of this app? (3 marks)

(6) You are given that F= $\dfrac{9}{5}$ C +32. Make C the subject of the formula (3 marks)

(7) Solve the equation $\sqrt{6 + \dfrac{x-3}{5x+2}} = 3$ (4 marks)

(8) y is inversely proportional to x^2. When x = 7, y = 4. Find the constant of proportionality hence find the value of y when x = 12   (3 marks)

(9) Simplify $y^{\frac{2}{3}} (y^{\frac{-2}{5}} + y^{\frac{1}{3}})$ (3 marks)

Answers to calculator based practice questions

(1) Answer: Constant of proportionality $=\frac{1}{2}$, and m = 2.5

(2) Answer: $x=\frac{1}{3}$, $x=-1\frac{1}{2}$

(3) Answer: x = 1 and $y = \frac{1}{2}$

(4) Answer x = 2.56 or 2.57 acceptable

(5) Answer: £5

(6) Answer: $C = \frac{5(F-32)}{9}$

(7) Answer: $x = \frac{-9}{14}$

(8) Answer: Constant of proportionality = 196, $y = \frac{196}{144}$ which simplifies to $\frac{49}{36} = 1\frac{13}{36}$

(9) Answer: $y^{\frac{4}{15}} + y$

Transformation Geometry

There are four basic transformations you need to know are:

(1) Reflection
(2) Rotation
(3) Translation
(4) Enlargements

Reflection

This is simply reflecting a shape about the mirror line. In this case we are reflecting the object ABC in the mirror line x = 4

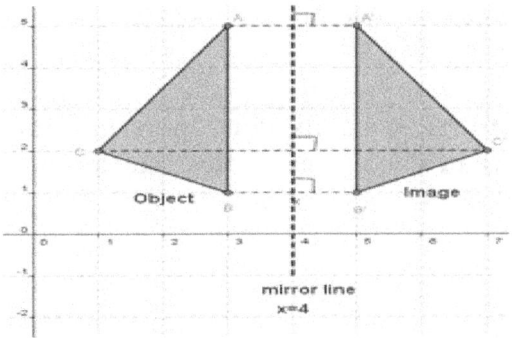

Rotation

A rotation is always anti – clockwise. So a rotation of say 200° of the lightly shaded triangle about a given centre has been rotated to its final place as shown by the darker triangle.°

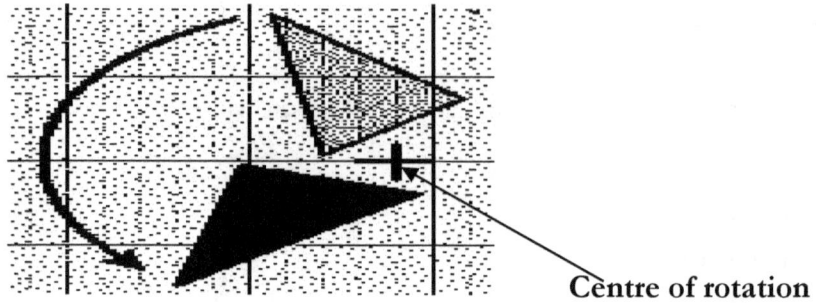

Centre of rotation

Important to Note: (1) A positive angle means the rotation is anti-clockwise and likewise a negative angle means it is a clockwise rotation about a given point.

(2) In this case the centre of rotation is marked '+'

Translation

This is simply moving a point or a shape along the x –axis and y-axis respectively.

For example the triangle ABC moves + 4 units along the x axis and + 3 units along the y – axis. We can also represent this translation by the translation vector $\begin{bmatrix} 4 \\ 3 \end{bmatrix}$

Y- axis

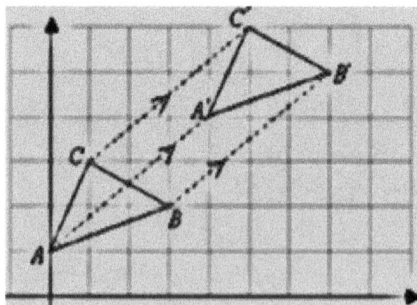

X - axis

Enlargements

Example: Below the triangle ABC is enlarged by scale factor +3 about the origin. The enlarged triangle is labelled A'B'C'. This means the triangle A'B'C' is 3× bigger than ABC and the centre of enlargement is the origin (0, 0)

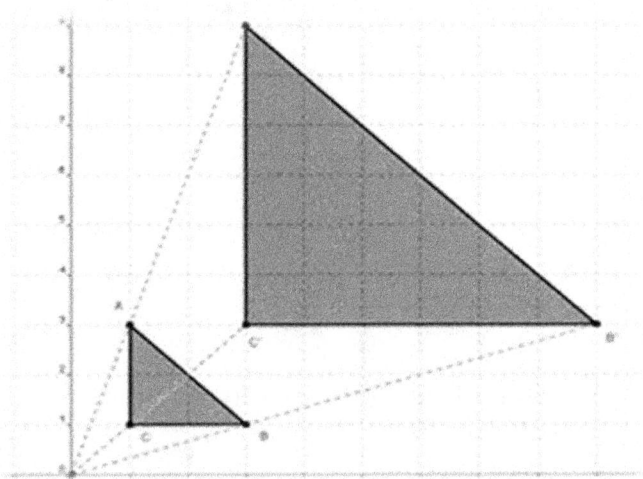

<u>Note:</u> **The length of OB'= 3×OB, OC' = 3×OC and OA' = 3×OA**

Geometry

Reminder of angles, triangles, parallel lines, common shapes & polygons:

(1) Angles on a straight line add up to 180 degrees

(2) The sum of the angles in a triangle add up to 180 degrees

(3) Right angled triangle – one angle in a triangle is 90°

(4) Equilateral Triangle - All sides and angles are equal. Each angle = 60°

(5) Isosceles triangle – base angles are equal and two sides are equal as shown below

(6) Exterior angle of a triangle

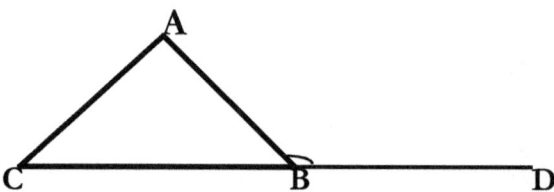

Exterior angle ABD = the sum of the opposite interior angles (Angle ABD = angle BAC plus angle ACB)

Parallel lines

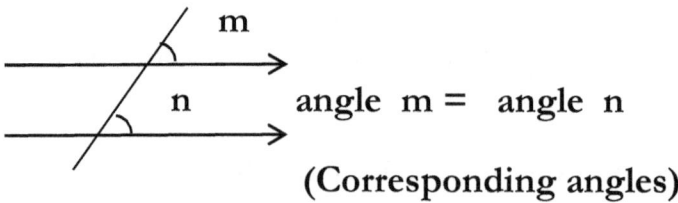

angle m = angle n

(Corresponding angles)

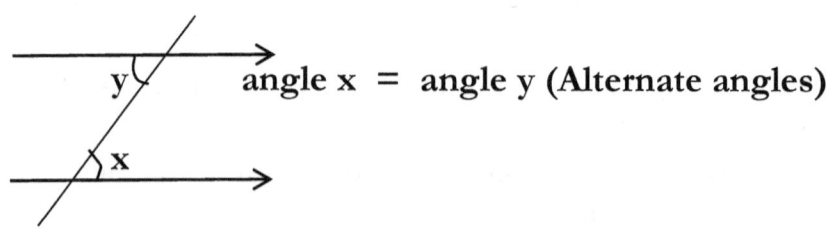

angle x = angle y (Alternate angles)

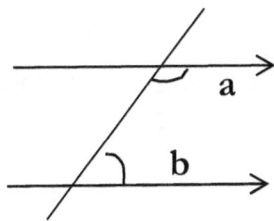 a + b = 180 degrees

Quadrilateral – 4 sided shape

The sum of the angles of a quadrilateral add up to <u>360 degrees</u>

Parallelogram   Trapezium

   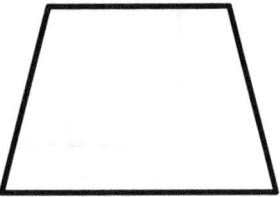

Rhombus (a squashed square)   Kite

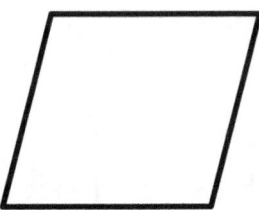   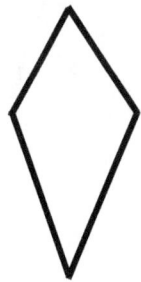

Regular Polygons

A regular polygon is where all the sides and angles are equal.

Consider these typical regular polygons:

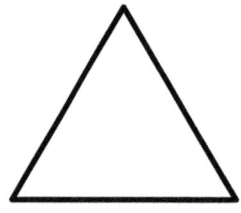

Regular 3 –sided polygon

Equilateral triangle (all sides equal)

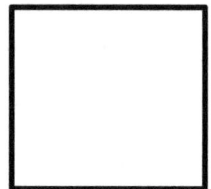

Regular 4 – sided polygon

Square

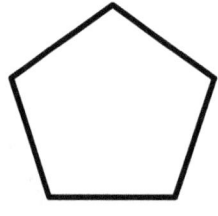

Regular Pentagon (5 sides)

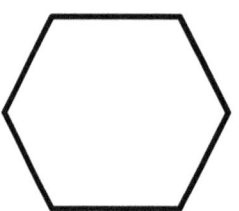

Regular Hexagon (6 sides)

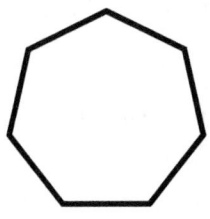 Regular Heptagon (7 sides)

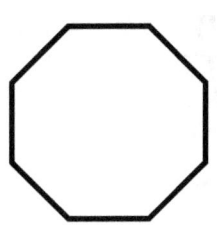 Regular Octagon (8 sides)

Useful formulas for polygons that you should memorize:

Exterior angle $= \dfrac{360}{n}$

Interior angle $= 180° -$ exterior angle $= 180 - \dfrac{360}{n}$ which simplifies to $\dfrac{180n-360}{n} = \dfrac{180(n-2)}{n}$

Angle based examples concerning triangles, parallelograms and polygons

Example 1: Find the interior angle x and the exterior angle y in the triangle below:

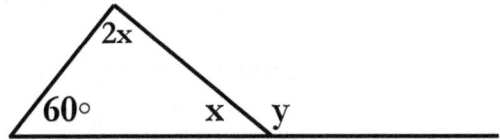

Method: 2x + x + 60 = 180 (interior angles of a triangle add up to 180°). This means 3x + 60 = 180, simplifying this we get 3x = 120. This means x = 40°.

To find y we can use the fact that x + y = 180. This means 40 + y = 180, hence y = 140°. So x = 40° and y = 140°

Example 2:

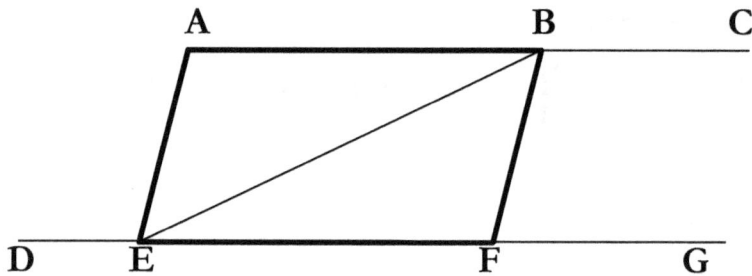

ABFE is a parallelogram. Angle ABE = 30° and angle BFG = 70°

Calculate angle AEB.

Method: Clearly angle BFE = 180 − 70 = 110° Also angle BEF = 30° (Alternate angle to angle ABE).

Now consider triangle BFE. Since we know two angles (angle BEF and angle BFE) we can work out the value of angle EBF. Angle EBF = 180 − (110 + 30) = 40°

Finally, angle AEB = angle EBF (alternate angles)

So angle AEB = 40°

Example 3: Find the interior and exterior angle of a regular octagon.

Using the formula for a regular polygon the interior angle is

$$\frac{180(n-2)}{n} = \frac{180(8-2)}{8} = \frac{180 \times 6}{8} = \frac{90 \times 6}{4} = \frac{90 \times 3}{2} = \frac{45 \times 3}{1} = 135°$$

Hence the exterior angle is 180 – 135 = 45°

Bearings

Bearings are measured clockwise from the **North line** to the line joining the two points as shown below:

Example: The bearing of A from B is measured clockwise from the North line and in this case is 75°

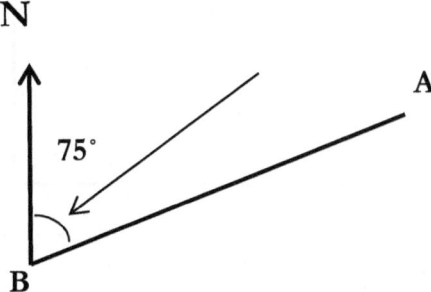

Example: Find the bearing of Q from P.

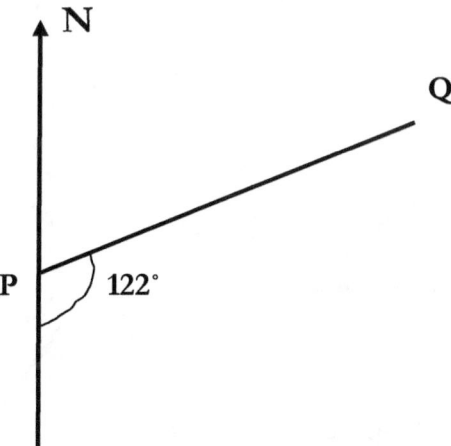

Method: Since you are going from P to Q we need to find the angle between the North line and Q.

This is 180° - 122° = **58°**

Areas and Volumes of common shapes

Perimeters, Areas and Volumes of common shapes

Consider the shapes below:

(1) Rectangle

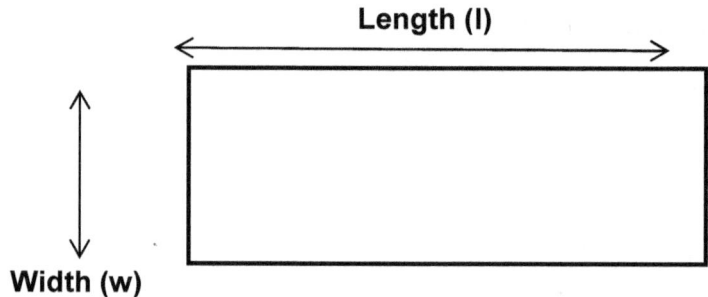

Area of a rectangle = Length X Width or $l \times w$

Perimeter of a rectangle = $2l + 2w$ (distance around the rectangle)

Note: Area is measured in units squared, e.g. cm^2 or m^2 and perimeter (distance all round a shape) is measured in the appropriate units e.g. cm or m

(2) Triangle

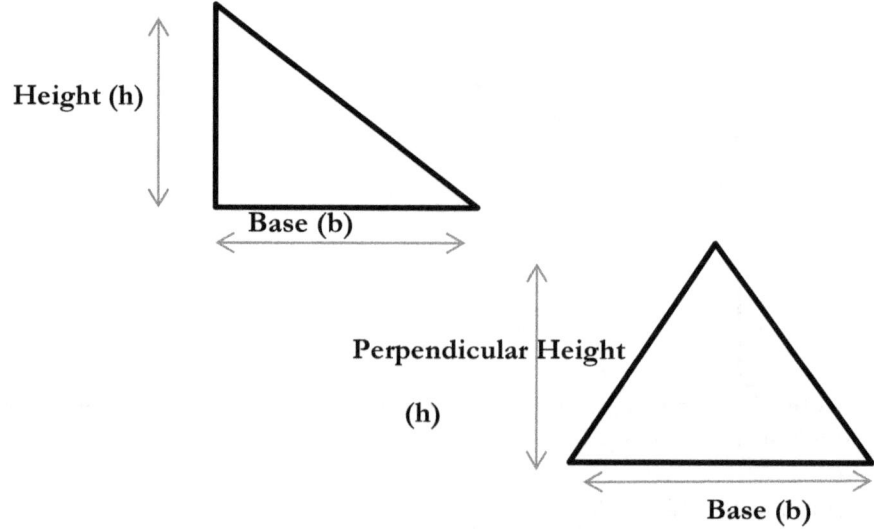

Area of a Triangle = 1/2 × base × height or $\frac{b \times h}{2}$ (The height is the perpendicular height relative to the base)

Area of a Trapezium

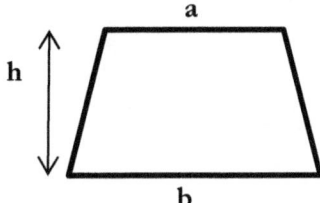

Area of a trapezium

Is equal to half the sum of the parallel sides × perpendicular height) = $\frac{(a+b)h}{2}$

Area of a circle is πr^2 (this means the value of π(pi) multiplied by radius squared)

Circumference of a circle (distance all the way round a circle) = $2\pi r$ or πd.

Note: **Diameter of a circle** = 2 × Radius

Approximate value of π = 3.142

Volume of a cuboid (or a box)

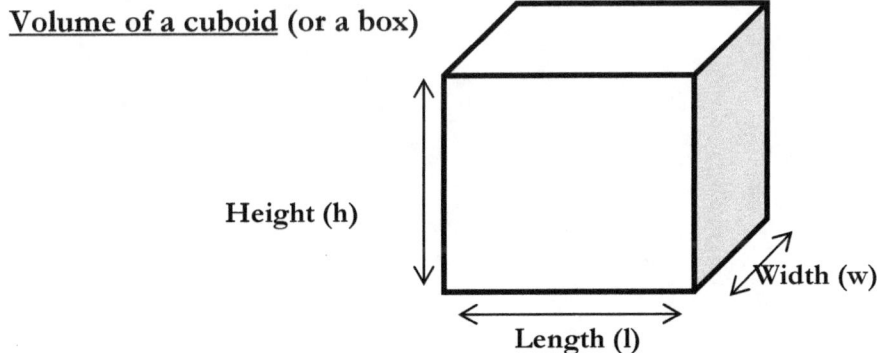

Volume of a cuboid is Height × Length × Width or V = h×l×w (units cubed e.g. cm^3 or m^3, etc)

Loci

A **locus** is simply the path taken by a point moving according to a prescribed rule. The plural of locus is loci.

Example 1: Bisecting a line segment

Using a compass draw the arcs shown from points X and Y

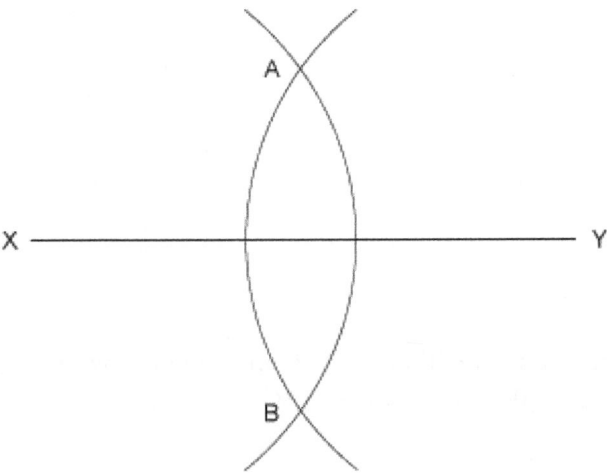

We can find the bisector of the line segment XY by joining A and B. The mid-point of XY is where the line AB intersects it as shown by the arrow:

Mid-point of XY is here:

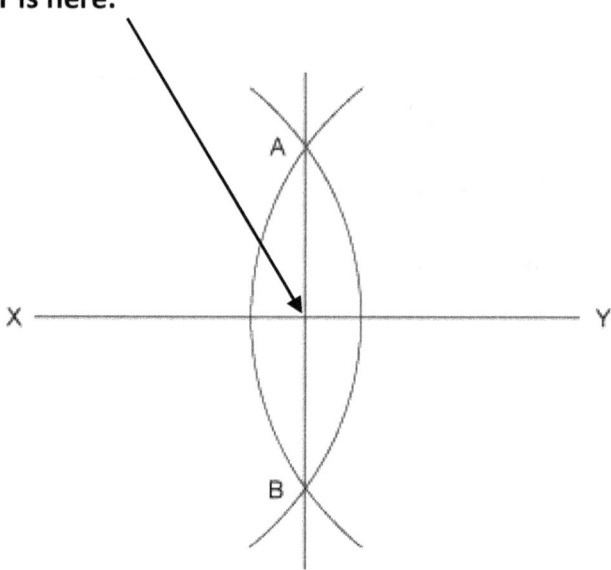

Bisecting an angle. To bisect an angle we follow the steps shown below: **Example**: Bisect angle ABC.

Given.

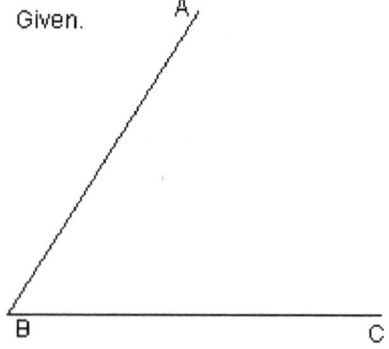

Step1: Draw an arc from point B as shown. Although this arc can have a radius of any length it has to intersect both sides of the angle. Let us call the intersection points P and Q

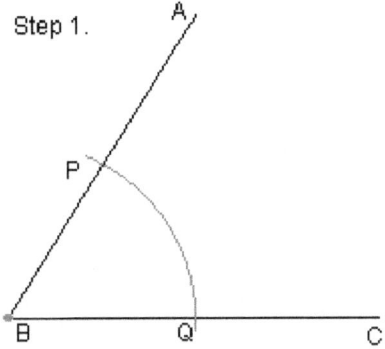

Step 2: From P and Q draw two more arcs. The radius for both the arcs drawn MUST be the same. Ensure that the arcs are long enough so that these two arcs intersect in at least one point. Let us call this intersection point X.

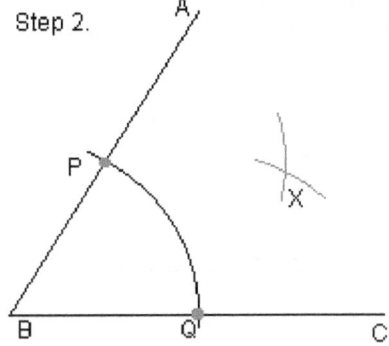

Step 3: Draw a line from B through X. The line which passes through points BX is the angle bisector.

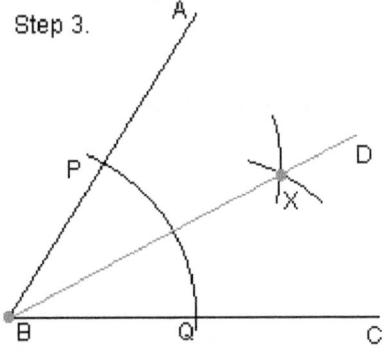

Circle Theorem

Firstly a few reminders:

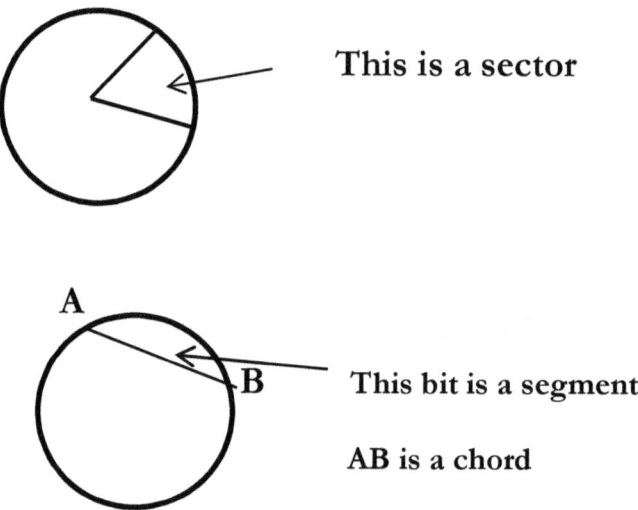

Basic rules of the circle theorem

(1) Angle subtended at the centre = 2 times the angle at the circumference.

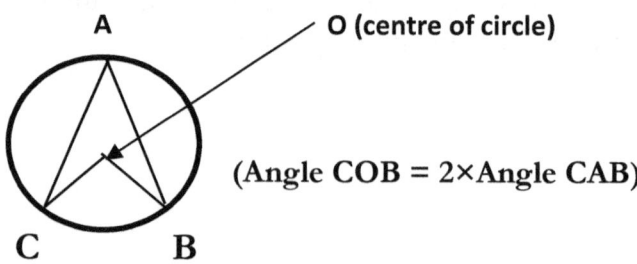

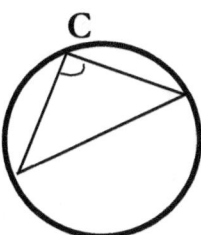

(2) Angle in a semi-circle is 90°

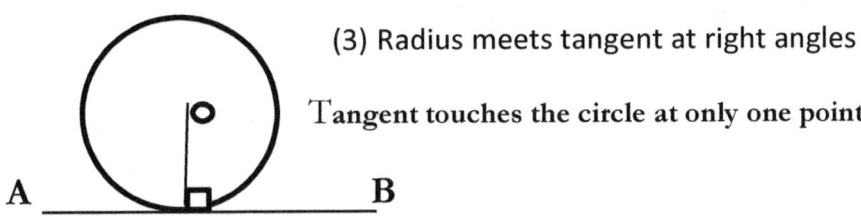

(3) Radius meets tangent at right angles

Tangent touches the circle at only one point

(4) The angle between the tangent and chord = angle in the alternate segment

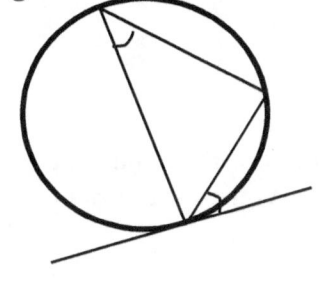

(The two angles shown are equal)

(5) The opposite angles of a cyclic quadrilateral add up to 180°

A cyclic quadrilateral is one that that touches the

circumference at each point as shown

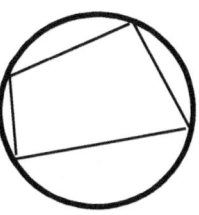

Example 1: Find the angle at the circumference denoted by A if the angle at the centre at O = 120°

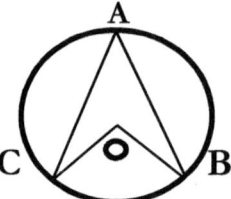

Method: Using the rule that the angle at the centre = 2 times the angle at the circumference, we can work out that the angle at A = half of angle COB. This means that the angle at A = 60°

Example 2: If angle at B = (x + y)°, what is the angle at A?

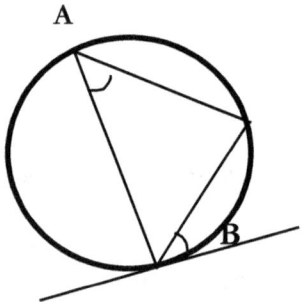

Method: Since the angle between the tangent and the chord (angle B) is equal to the angle in the alternate segment (angle A) then it follows that angle A also = (x + y)°

Area of a Sector of a circle

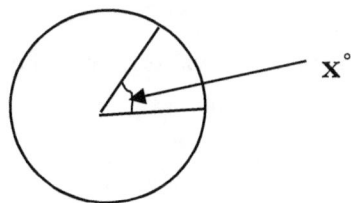

$$\text{Area} = \frac{x}{360} \times \pi r^2$$

Example: Find the area of a sector whose angle at the centre is 60° and the radius of the circle is 10cm. Write the answer to two decimal places

Method:

$A = \frac{x}{360} \times \pi r^2$

$A = \frac{60}{360} \times \pi r^2 \quad A = 0.167 \times \pi r^2$

$A = 0.167 \times 3.142 \times 100 \implies A = 52.5\ cm^2$

Linear Equations

These are of the form y = mx + c

Where m is the gradient or slope and c is the value of y when x = 0

Example 1: y = 3x - 1

The graph is shown below.

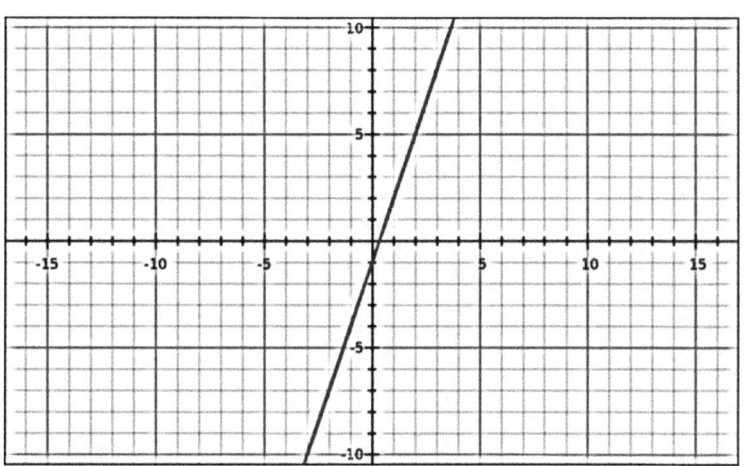

You can see that the equation y = 3x – 1 crosses the y –axis at y = -1 (this is the called the intercept)

In other words when x = 0, y = 3×0 – 1 = -1

The '3' in the 3x bit refers to the gradient or the slope of the graph.

So in general a linear equation is of the form y = mx + c, where m is the gradient and c is the value of y when x = 0

Example 1

Plot the equation y = 2x - 3 for values of x = -2 to +2 by completing the table below first. The plotted graph is shown below the completed table.

x	-2	-1	0	1	2
2x – 3	-7	-5	-3	-1	1
y	-7	-5	-3	-1	1

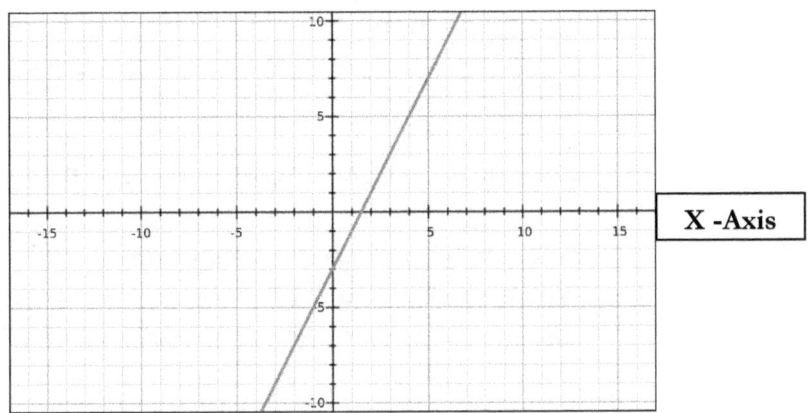

Example 2

The co-ordinates P (2, 3) and Q (4, 6) lie on a straight line.

  (i)   Find the mid- point R of the co-ordinates P & Q
  (ii) Find the gradient of the straight line
  (iii) Find the equation of the straight line

(i) The mid-points are simply the average of the x – coordinates and the y – co-ordinates.

Mid –point for the x co-ordinate = $\frac{x_1+x_2}{2} = \frac{2+4}{2} = 3$

Similarly, the mid-point of the y- co=ordinate = $\frac{y_1+y_2}{2} = \frac{3+6}{2} = 4.5$

Hence R, the mid-point of PQ is (3, 4.5)

(ii) The gradient of two points that lie on a straight line = $\frac{Difference\ in\ y-co-ordinates}{Difference\ in\ x-co-ordinates} = \frac{6-3}{4-2} = \frac{3}{2} = 1.5$

(iii) The equation of a straight line is given by y = mx + c

Since m = 1.5, then the equation is y = 1.5x + c
We also know that it goes through P, Q and R. So we can choose any of these to find the value of C. Let us choose P(2, 3) so the equation is now:
3 = 1.5×2 + C, Hence 3 = 3 + C so in this case C = 0
So the equation is y = 1.5x

Graphs

Quadratic Equations

These are of the form $f(x) = ax^2 + bx + c$. Note $f(x)$ is called a function of x since y is defined in terms of x.

Consider the example below:

Plot the equation $y = 3x^2 - 2x + 1$

First choose suitable values of x say from -3 to +3 and find the corresponding values of y as shown in the table below:

x	-3	-2	-1	0	1	2	3
$3x^2$	27	12	3	0	3	12	27
-2x	6	4	2	0	-2	-4	-6
+1	1	1	1	1	1	1	1
Y	34	17	6	1	2	9	22

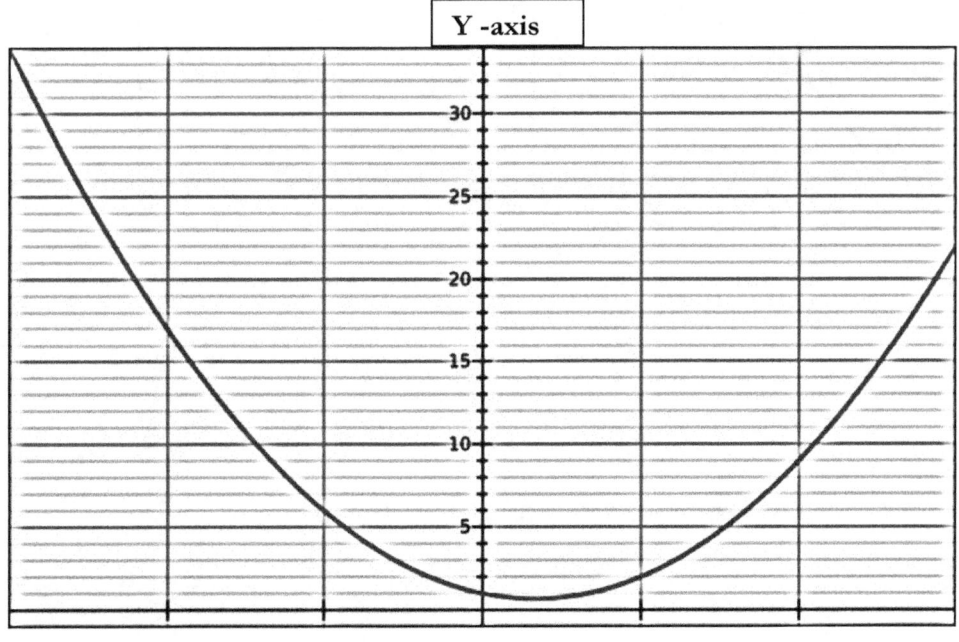

Cubic equation

Example: Plot the equation $y = x^3 - 1$

x	-3	-2	-1	0	1	2	3
x^3	-27	-8	-1	0	1	8	27
-1	-1	-1	-1	-1	-1	-1	-1
Y	-28	-9	-2	-1	0	7	26

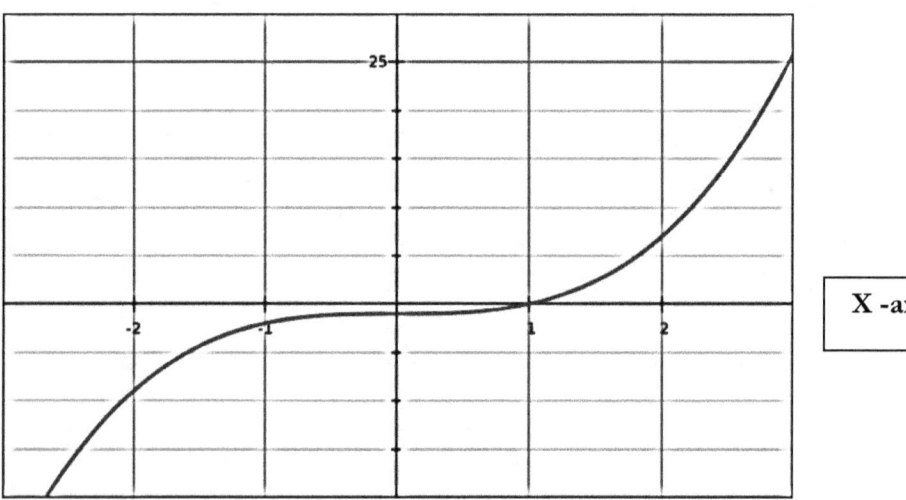

Solving equations using graphical methods

Example

You are given that the quadratic equation $y = x^2 - 4x + 8$ and the linear equation $y = 3x - 2$ intersect at two points A and B. Find the co-ordinates of these two points A and B where the equations intersect.

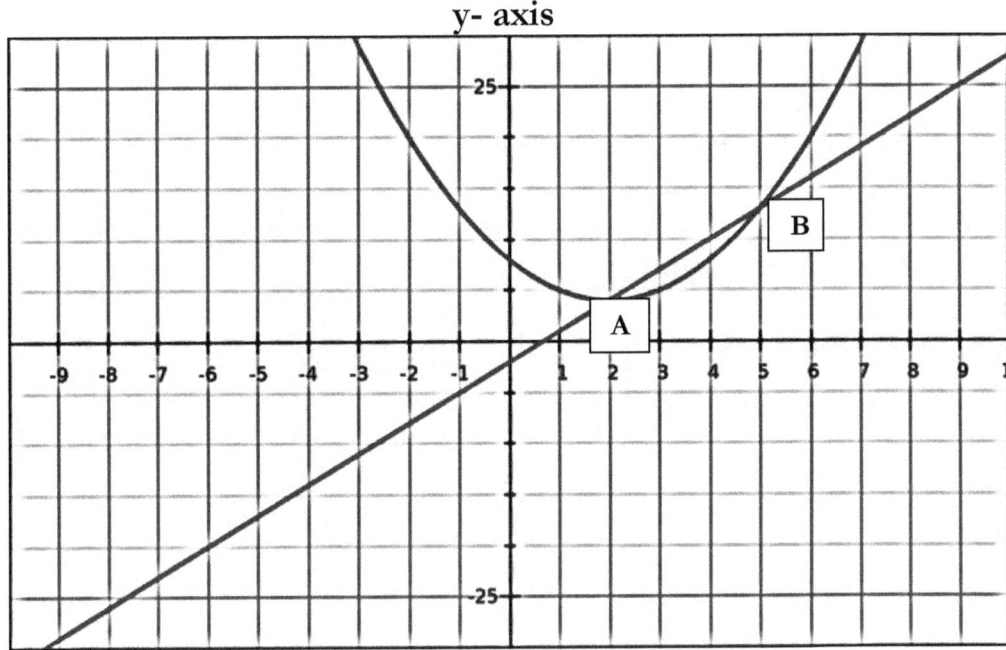

Method: First plot the equations $y = x^2 - 4x + 8$ and the linear equation $y = 3x - 2$ as shown above.

You can see that the co-ordinates of A are (2, 4) and the co-ordinates of B are (5, 13)

You can of course solve these simultaneous pair of equations mathematically as shown below:

The two equations are $y = x^2 - 4x + 8$ and $y = 3x - 2$.

This means $x^2 - 4x + 8 = 3x - 2$

Simplifying this we get $x^2 - 7x + 10 = 0$ ⟹ $(x-5)(x-2) = 0$

This means either, $x - 5 = 0$ or $x - 2 = 0$ ⟹ $x = 5$ or 2. We can now find the corresponding values of y by substituting these values in the equation $y = 3x - 2$

When $x = 5$, $y = 3 \times 5 - 2 = 13$ and when $x = 2$, $y = 3 \times 2 - 2 = 4$

So the co-ordinates of A = (2, 4) and B = (5, 13)

Transformations of functions

When $y = f(x + a)$ the graph moves 'a' units left. (It is opposite of what you might expect)

Likewise when $y = f(x - a)$ the graph moves 'a' units to the right

Consider the two graphs below (1) f(x) and (2) f(x) = f(x +2)

(1) $f(x) = x^2 + 2$ and   (2) $f(x+2) = (x+2)^2 + 2$ shown below

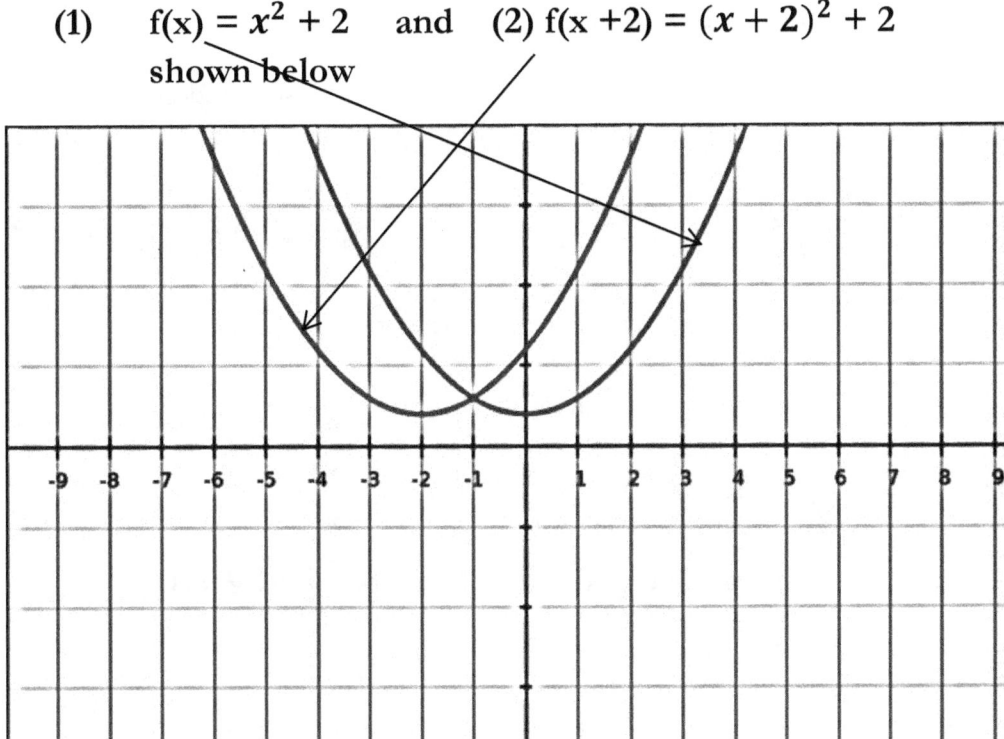

You can see that f(x + 2) has shifted to the left by 2 units (this seems counter-intuitive) but f(x + 2) does shift to the left by 2 units and <u>not</u> to the right.

Other types of transformations involving y = f(x)

It is worth remembering that y = f(x) + a moves up the y-axis by 'a' units and likewise y = f(x) – a moves down the y –axis by 'a' units.

Finally y = k×f(x) or kf(x) simply means the graph of f(x) stretches along the y –axis by a factor of k.

Inverse and Composite functions

Inverse Functions:

Example 1: If $f(x) = 3x + 2$ find the inverse function $f^{-1}(x)$

Method:

Since $f(x) = 3x + 2$, we can re –write this as $y = 3x + 2$.

Now make x the subject. So we have $y - 2 = 3x$ or $x = \frac{y-2}{3}$

This means the in inverse function can be written as:

$$f^{-1}(x) = \frac{x-2}{3}$$

Example 2: If $g(x) = 5x - 7$ find its inverse $g^{-1}(x)$

As before we can re-write If $g(x) = 5x - 7$ as $y = 5x - 7$.

We now make x the subject. This means $y + 7 = 5x$

Or $x = = \frac{y+7}{5}$

So $g^{-1}(x) = \frac{x+7}{5}$

Composite functions:

When functions are combined it becomes a composite function. If you have two functions f(x) and g(x). The composite function fg(x) simply means you need to g first and then f.

Example: You are given that $f(x) = 3x - 1$ and $g(x) = x^2$

(a) Work out f(g(3)) when x= 3 and (b) gf(4) when x = 3

(a) The order in which you do this is important. So to work out f(g(3)) we first work out g(3). Now g(3) = 3^2 = 9, hence, f(9) = 3×9 − 1 = 27 − 1 = 26

(b) Similarly to work out g(f(x)), we first work out f(3) = 3×3 − 1 = 8. This means g(8) = 8^2 = 64. So g(f(x)) = 64

Plan and elevation

An example of this is when designing buildings you need to show 2D drawings to show what the building will look like from each side.

The key things to remember is:

(1) The view from the top is called the plan.
(2) The views from the front and sides are called the elevations.

> **Example:** Draw the plan and elevation of the cube. **Method:** In this case both the plan and the elevation is shown as in the drawing shown.

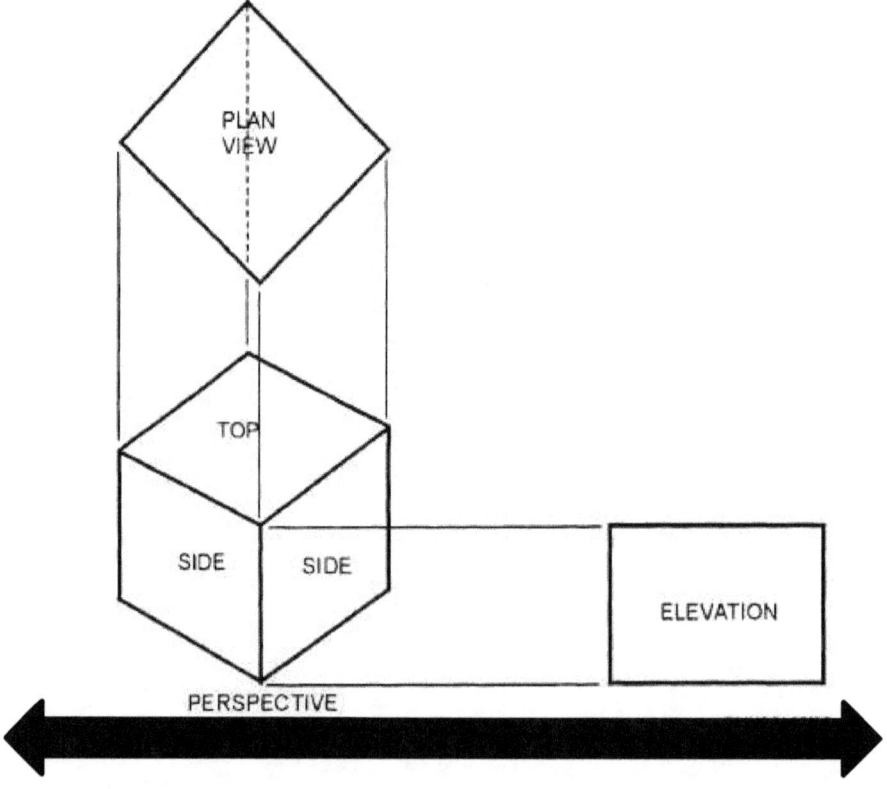

Trigonometry for Right Angled Triangles

Formulae for Right Angled Triangle

In a right-angled triangle you need to know:

The Sine of an angle = the ratio of the **Opposite Side** to the **Hypotenuse**

The Cosine of an angle = the ratio **of Adjacent Side** to **Hypotenuse**

The Tangent of an angle = the ratio of the **Opposite Side** to the **Adjacent Side**

You can also try remembering S**OH-CAH-TOA**

SOH –   SIN (ANGLE) = **OPPOSITE SIDE/HYPOTENUSE**

CAH –   COS (ANGLE) = **ADJACENT SIDE/HYPOTENUSE**

TOA –   TAN (ANGLE) = **OPPOSITE SIDE/ADJACENT**

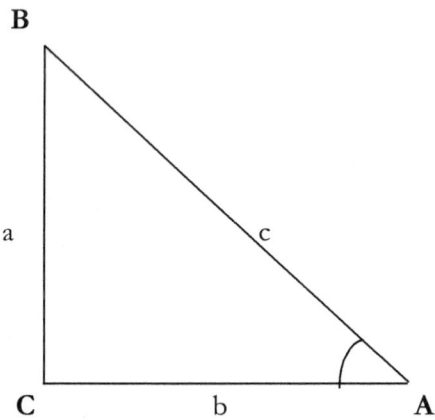

$$\text{Sin (A)} = \frac{a}{c}$$

$$\text{Cos (A)} = \frac{b}{c}$$

$$\text{Tan (A)} = \frac{a}{b}$$

Trigonometry for non- right angled triangles

Formula for a <u>non-right</u> angled triangle are shown below

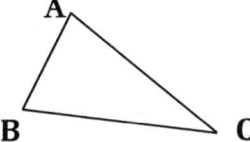

Sine Rule:

$$\frac{a}{SinA} = \frac{b}{SinB} = \frac{c}{SinC}$$

Cosine Rule:

$$a^2 = b^2 + c^2 - 2bcCosB$$

$$b^2 = a^2 + c^2 - 2acCosB$$

$$c^2 = a^2 + b^2 - 2abCosC$$

(**Note**: Although there are three versions of the formula they all have the same pattern)

Example1:   In the triangle below find angle B

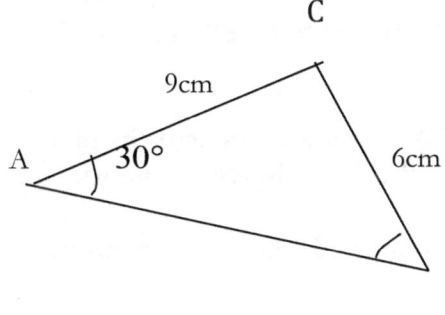

Using the Sine rule: $\dfrac{a}{SinA} = \dfrac{b}{SinB}$ substitute the appropriate known angles and sides in the formula shown

$$\dfrac{6}{Sin30} = \dfrac{9}{SinB}$$

Hence, Sin B = 9 X Sin30/6 = 0.75. So the angle B = 48.6°

Example 2: In the triangle shown find the angle C

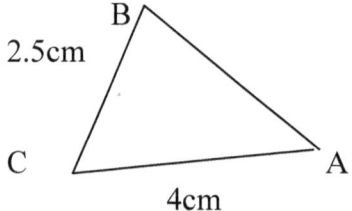

Using the Cosine Rule, we can say that $c^2 = a^2 + b^2 - 2abCosC$

Hence, $CosC = (a^2 + b^2 - c^2)/2ab$ = (9 + 16 – 6.25)/24 = 18.75/24 = 0.78125

Hence, C = 38.6°

Summary for using the sine rule and Cosine rule

You can use the sine rule when: (1) Two angles are given and a side or (2) two sides are given as well as an angle <u>not enclosed by them</u>

You can use the cosine rule when: (1) You know the lengths of 3 sides Or (2) you know two sides and the angle between them

Also note that the area of a non- right angled triangle is given by $\frac{1}{2}abSinC$

Graphs of Trigonometrical Functions

Consider graphs of y = sin(x) and cos(x)

For y=sin(x) maximum value of y = 1 when x =90° and minimum value of y = - 1 when x= -90°. This cycle repeats every 360°

Also note that when x = 0°, sin(0) =0, and when x = 360°, sin(360) = 0

Similarly for y = cos(x). The maximum value of y = 1, when x = 0° and minimum value of y = -1 when x = 180°. Again this cycle repeats every 360°

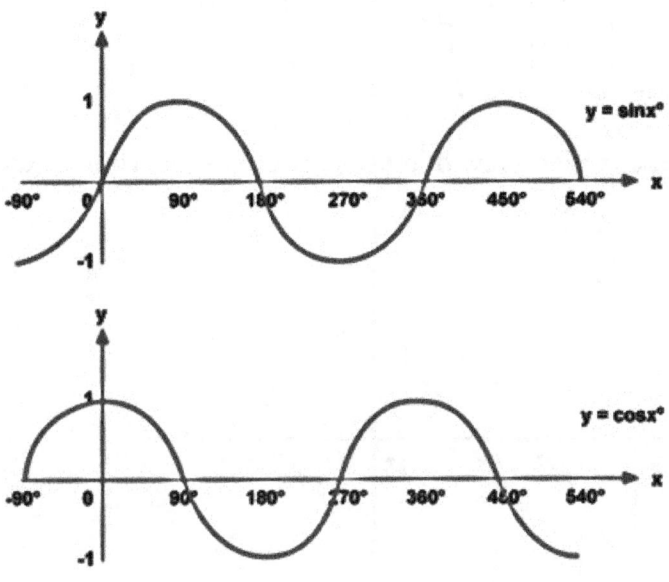

y= tan(x)

The values of tan (x) repeat every 180°. As x approaches 90°, the value of y or tan(x) approaches infinity. The symbol for infinity is ∞.

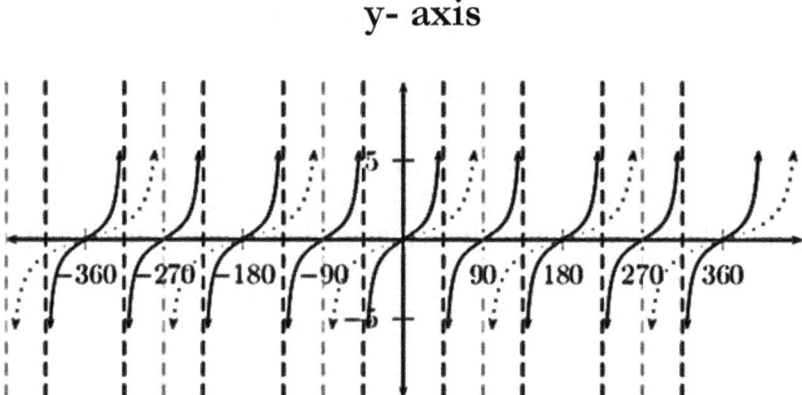

Below is a magnified view of a tangent curve this time of y= tan(θ).

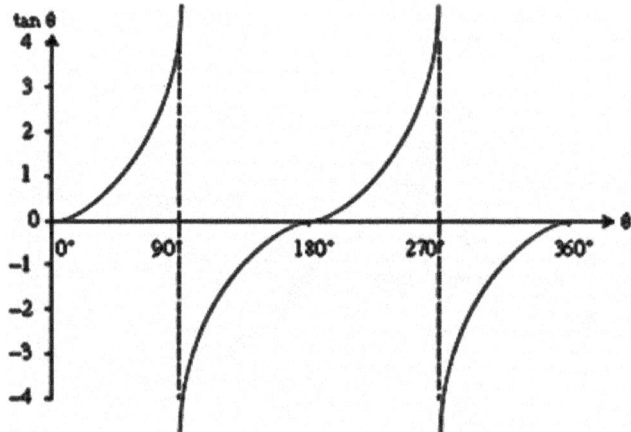

You can see that as θ approaches 90°, tan (θ) tends to infinity

Similarly, as θ approaches -90°, tan (θ) tends to minus infinity

Solving trigonometric equations using graphical methods: Solve the equation sin(x) = 0.5 for values of x between 0 and 360°

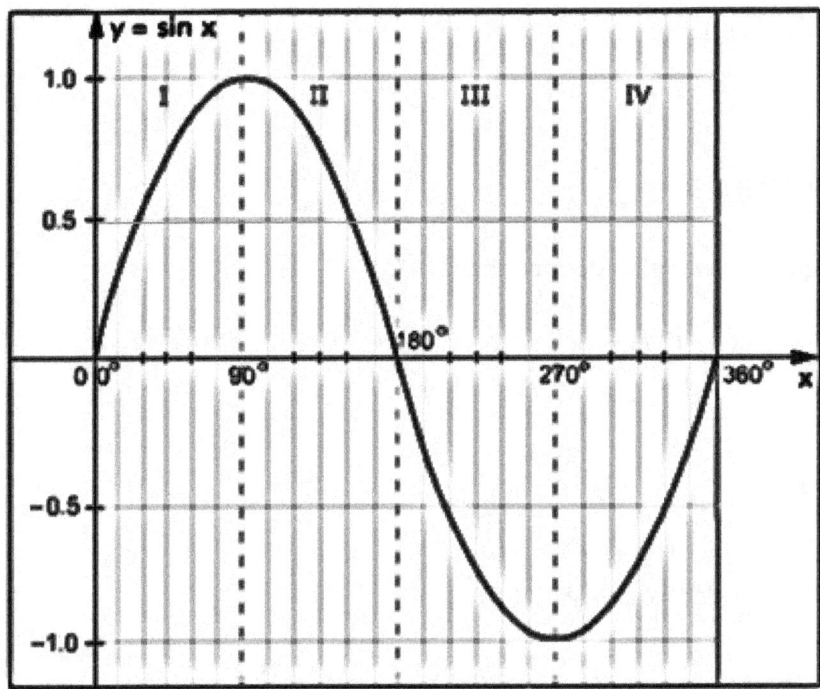

From the graph you can see that when y = 0.5, it meets the curve when x = 30° and 150°

Trig Identities

Rule 1

$\sin^2 x + \cos^2 x = 1$

Re-arranging we can also deduce that $\sin^2 x = 1 - \cos^2 x$

Similarly, $\cos^2 x = 1 - \sin^2 x$

Rule 2

$\text{Tan } x = \frac{\sin x}{\cos x}$

Example 1: Simplify $\sin^2 x + \cos^2 x + \frac{\sin x}{\cos x} - 1$

Method: We know that $\sin^2 x + \cos^2 x = 1$ and $\frac{\sin x}{\cos x} = \tan x$

➡ $\sin^2 x + \cos^2 x + \frac{\sin x}{\cos x} - 1 = 1 + \tan x - 1 = \tan x$

Example 2: If $\sin x = \frac{1}{2}$, Find the value of cos x in surd form

Method: Using $\sin^2 x + \cos^2 x = 1$ we can deduce that $\frac{1}{2} \times \frac{1}{2} + \cos^2 x = 1$

➡ $\frac{1}{4} + \cos^2 x = 1$ ➡ $\cos^2 x = \frac{3}{4}$ ➡ $\cos x = \frac{\sqrt{3}}{2}$

Example 3: Solve the equation 4sinx + 6 = 9 for values of $0° \leq x \leq 90°$ (Give your answer to 1 d.p)

Simplifying the equation 4sinx + 6 = 9, we get 4sinx = 3 ➡ $\sin x = \frac{3}{4}$ = 0.75. Hence x = 48.6°

Similarity and Congruence

Similarity

Similar figures have the same shape but different sizes.

Example 1:

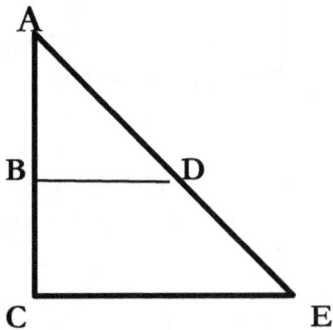

<u>Triangle ABD and AEC are the same shape but are different sizes.</u>

In this case we can use ratios of sides to determine one of the unknown sides. For example we can deduce that $\frac{AB}{AC} = \frac{BD}{CE}$, so if we know the values of AB, AC and CE we can determine the value of BD

Things to notice in similar triangles:

(1) Corresponding angles are equal (angle BAD = angle CAE)

(2) Corresponding lengths are in the same ratio (e.g. $\frac{AB}{AC} = \frac{BD}{CE}$)

Example 2

Consider the two cylinders shown which are mathematically similar.

Radius = 4cm

Height = 9cm

Radius = 12cm

Height = 27cm

Notice the corresponding lengths are of the same ratio. Radius of bigger cylinder is three times the smaller cylinder. Also the height of the bigger cylinder is three times that of the smaller cylinder.

Points to note: In similar figures although the corresponding lengths are in the same ratio the areas and volumes are not in the same ratio as their lengths.

Consider two similar rectangles shown below:

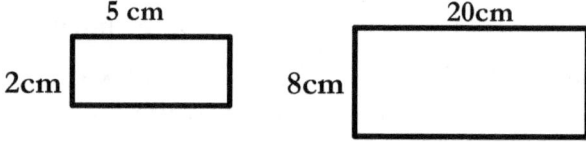

In the above similar rectangles as we noted before the ratio of the lengths are the same that 1:4. But now consider the areas. Area of small rectangle $= 5\times2 = 10cm^2$

And the area of the larger rectangle $= 8\times20 = 160cm^2$. The ratio of the 'areas' of these similar rectangles are 1: 16. Put another way if the shape is enlarged by scale factor +4, then the area is bigger by 4^2, Similarly, the ratio of the 'volumes' of similar shapes correspond to the cube of the lengths.

Example: The two cubes shown have lengths that are as follows: lengths of small cube are 2cm by 2cm by 2cm. And the lengths of a similar cube are enlarged by scale factor 2. What is the volume of the larger cube?

Clearly the volume will be 8× the volume of the smaller cube (since $2^3 = 8$). So the volume of the larger cube is $8 \times 8 = 64 \ cm^3$ (Proof: Second cube is 2× the length of the first cube. This means the dimensions of the second cube are 4cm by 4cm by 4cm. Hence the volume of the larger cube is $4 \times 4 \times 4 = 16 \times 4 = 64 cm^3$

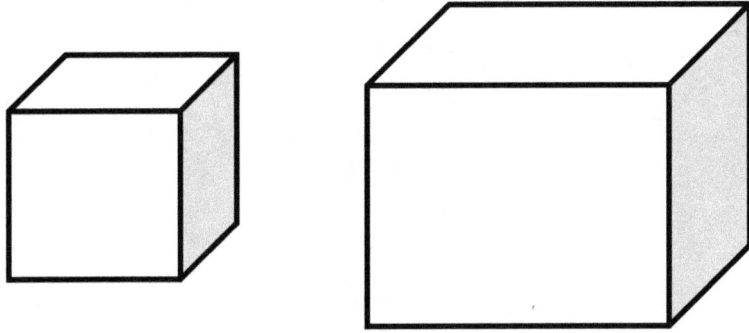

Example: Consider the two similar cans below.

Assume the volume of liquid that is held by the smaller can is 0.8 liters what is the volume that could be held by the larger can?

Height = 5 cm   **Height = 12cm**

Solution: The ratio of the bigger can to the smaller can $= \frac{12}{5} = 2.4$

Hence the volume of that could be held by the bigger can is $2.4^3 \times 0.8 = 11.0592$ litres or 11.1 litres to 1 d.p.

Congruent Triangles

Two triangles are said to be congruent if certain set of rules are true.

1. Three sides of both the triangles are same (SSS)
2. Two sides and the angle between them in one triangle = to two sides and the included angle in the other triangle (SAS)
3. If each triangle contains a right angle. The hypotenuse and another pair of sides are equal (RHS)
4. Two angles and a side in one triangle = two angles and the corresponding side in the other triangle (AAS)

Pythagoras' theorem

All you need to remember for this is the formula as shown below.

(Note: This theorem is only true for right angled triangles)

The square of the hypotenuse = the sum of the squares of the other two **sides**.

$$h^2 = a^2 + b^2$$

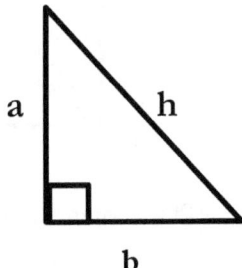

Example 1:

In the triangle below calculate the value of the side b.

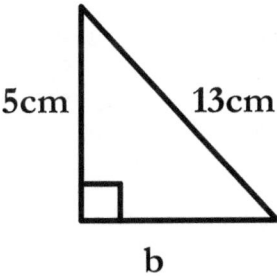

Method: Using Pythagoras' theorem we have $h^2 = a^2 + b^2$

Substituting the values we have $13^2 = 5^2 + b^2$

$\Rightarrow$ $169 = 25 + b^2$ $\Rightarrow$ $169 - 25 = b^2$ $\Rightarrow$ $144 = b^2$

Hence $b = \sqrt{144} = 12$

Example 2:

You are given that in the diagram below: HG = 12 cm, GE = 8cm, BE = 6cm. Find HB. Give your answer to 4 S.F.

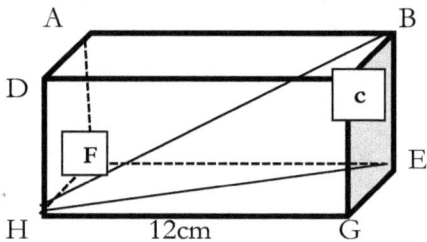

Method: First find HE using Pythagoras' theorem $HE^2 = GE^2 + HG^2$ (Since we know HG & GE)

➡ $HE^2 = 12^2 + 8^2 = 144 + 64 = 208$ ➡ $HE = \sqrt{208}$

We can now find HB, since $HB^2 = HE^2 + BE^2$ ➡ $HB^2 = 208 + 36 = 244$

➡ $HB = \sqrt{244} = 15.62$ cm

Volumes and Surface Areas

You probably already know that the volume of a cuboid is its length×width×height in units cubed. That is V = l×w×h = lwh units cubed

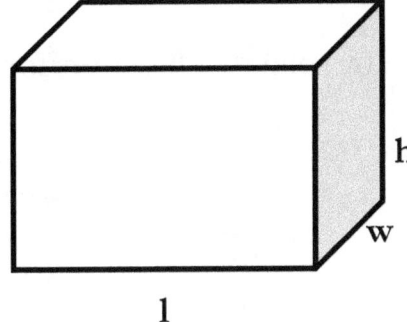

The surface area of the cube = 2(lw + hw + lh)

You need to also need to know the volume of other 3-D shapes.

For example the volume of a cylinder or any prism is its **base area× height**

The volume of a cylinder = $\pi r^2 h$ (where r is the radius and h is its height)

The surface area of a cylinder = $2\pi rh + 2\pi r^2$

Example 1: Find the volume of the cylinder shown whose radius is 10cm and height is 15cm

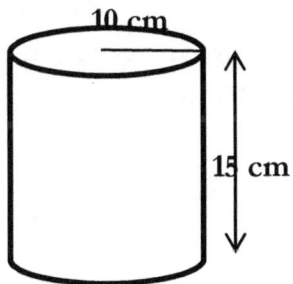

Method: Volume = Base area ×height = area of circle × height = π×10×10×15 = 3.142×100×15 =314.2×15 = 4713 cm^3

Example 2: Find the volume of the triangular prism below. Whose height is 8cm, the width is 6cm and length is 9 cm

(Note: A prism is any solid that has a **uniform cross – section)**

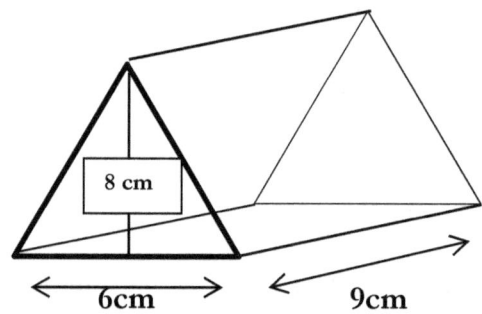

Method: Volume of the triangular prism is its triangle base area × length = $\frac{1}{2}$×6×8×9 =216 cm^3

Volume of a pyramid or a cone

Volume of a pyramid = $\frac{1}{3}$ base area ×height

Volume of a cone also = $\frac{1}{3}$ base area ×height

Surface Area of a cone = area of circular base + curved area of cone = πr^2+πrl (where r is the radius and l is the slant length)

Example 3:

Find the volume of the pyramid shown below. It has a height of 11 metres and a square base whose sides are 6m each.

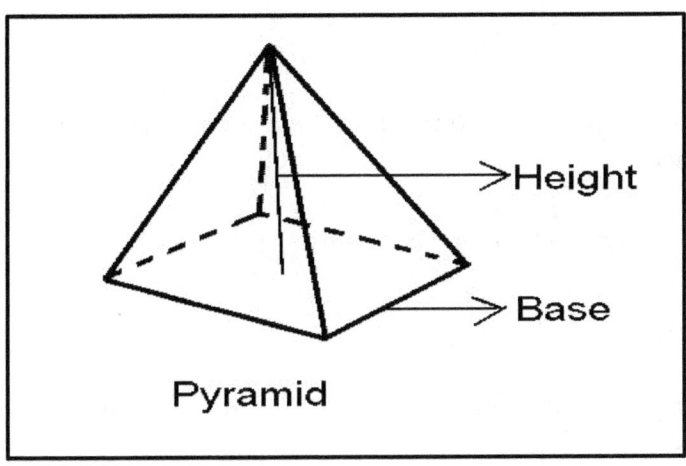

Step 1: Volume of a pyramid = $\frac{1}{3}$ base area ×height

Step 2: Volume = $\frac{1}{3}$ ×6 ×6×11 = 2 ×6 ×11 =12×11 = 132 m^3

Example 4: Find the volume of the cone below whose radius is 5cm and the height is 12 cm. Give your answer to 4 significant figures.

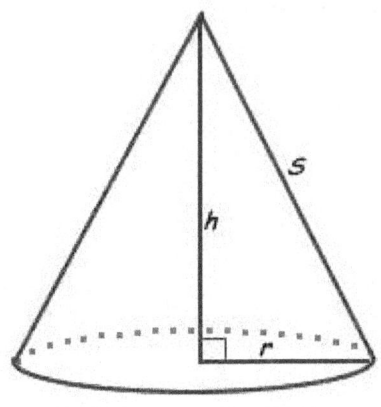

Volume of cone = $\frac{1}{3}\pi r^2 h$

Hence Volume = = $\frac{1}{3} \times \pi \times 5^2 \times 12 = 314.2 \ cm^3$

Another volume question that may come up in the exam is to work out the volume of a sphere:

Volume of a sphere = $\frac{4}{3}\pi r^3$ (where r is the radius)

Example 5:

The volume of a cylinder is $3000 cm^3$. Its radius is 6.5cm. Calculate the height of the cylinder to 3 S.F.

Method: Step 1: Volume of cylinder = $\pi r^2 h$, so $\pi r^2 h = 3000 cm^3$

Step 2: Substitute the value of π and r and make h the subject

$3.142 \times 6.5^2 \times h = 3000$, h = $\frac{3000}{3.142 \times 42.25}$ = $22.59896 \ cm^3$ = $22.6 cm^3$ to 3 S.F.

It is also useful to know the volume and surface area of a sphere.

Volume of a sphere = $\frac{4}{3}\pi r^3$ (where r is the radius)

So volume of a hemisphere = $\frac{2}{3}\pi r^3$

Surface area of a sphere = $4\pi r^2$

Vectors

Vectors have both magnitude and direction. For example wind has both magnitude and direction. So does velocity or acceleration.

Consider the vector $\begin{pmatrix} 3 \\ 4 \end{pmatrix}$

Vectors can be represented diagrammatically or simply by a letter e.g. 'a' or by $\overrightarrow{AB}$ Consider the vector $\begin{pmatrix} 3 \\ 4 \end{pmatrix}$ diagrammatically as shown below.

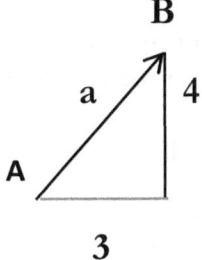

You can see that we can find AB or 'a' by using Pythagoras' theorem.

$a^2 = 3^2 + 4^2$ ⟹ $a^2 = 9 + 16$ ⟹ $a^2 = 25$ ⟹ $a = \sqrt{25} = 5$

Some rules you should know:

If $\overrightarrow{AB}$ = a then $\overrightarrow{BA}$ = -a

If $\overrightarrow{AB}$ = k$\overrightarrow{AB}$ ⟹ $\overrightarrow{AB}$ is parallel to K$\overrightarrow{AB}$

Vectors can be added, subtracted or multiplied

Example: If $a = \begin{pmatrix} 3 \\ 4 \end{pmatrix}$ and $b = \begin{pmatrix} 7 \\ 8 \end{pmatrix}$

Then $3a - b = \begin{pmatrix} 9 - 7 \\ 12 - 8 \end{pmatrix} = \begin{pmatrix} 2 \\ 4 \end{pmatrix}$

In a vector diagram the resultant can be found be found by the arrow facing in the opposite direction as shown below:

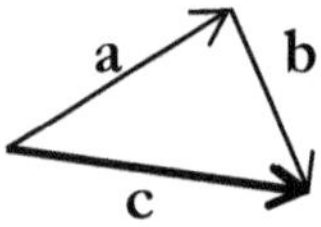

$c = a + b$

(Notice c points in the opposite direction to a and b)

If $u = \begin{bmatrix} x \\ y \end{bmatrix}$ then its magnitude, $\lfloor u \rfloor = \sqrt{x^2 + y^2}$

Example

In the vector diagram below find the vectors $\vec{AM}$ and $\vec{CM}$ in terms of a and b

You are given that $\vec{AM} = \frac{1}{4}$ of $\vec{AB}$. Also that $\vec{CA} = b$ and $\vec{BC} = a$

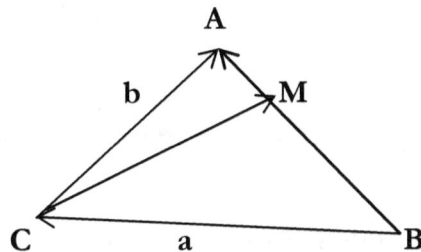

From the vector diagram we can deduce that $\vec{BC} + \vec{CA} = \vec{BA}$

$\Rightarrow$ $a + b = -\vec{AB}$ or $\vec{AB} = -a - b$ $\Rightarrow$ $\vec{AM} = \frac{-(a+b)}{4}$

To find $\vec{CM}$ we can say that $\vec{CM} = b + \frac{-(a+b)}{4} = \frac{4b-a-b}{4}$

Arithmetic Sequences

Working out a general formula for an arithmetic sequence:

You can either use the difference method or the general formula to find the nth term of an arithmetical sequence.

Example 1: (a) Find the nth term of the sequence 5, 8, 11, 14, ----- **(b)** Hence find the 15th term.

Method: The common difference in this case is 3. So we multiply the common difference by n to get 3n. However each term is 2 more than 3n. (a) Hence the nth term is 3n + 2

(b) To find the 15th term simply substitute n = 15 in 3n +2. Hence the 15th term is 3×15 + 2 = 45 + 2 = 47

Example 2: Find the nth term of the sequence 4, 1, -2, -5

Method: This time the common difference is -3. So the nth term is 7 – 3n

(Since each term is 7 more than -3n so the nth term is 7 – 3n

This time consider a general arithmetical sequence as shown:

a, a +d, a+2d, a +3d, a+4d, a+5d, a+6d, ... (d is the common difference). We can see that the second term is a+d, the third term is a+2d, the fourth term is a+3d, the fifth term is a+4d or a + (5-1)d, the sixth term is a+5d or a +(6-1)d

The seventh term is a+6d or a + (7- 1)d, so the nth term is a+(n-1)d

You can check to see if this is right by substituting n=1, 2, 3, 4, 5 and so on to the appropriate numbers in the sequence. See example below:

Example 2: Find the nth term of the arithmetical sequence below:

5, 9, 13, 17 ... this is an arithmetical or linear sequence since the numbers go up by the same constant number. We know the nth term is a + (n - 1)d

In this case a=5 (This is the first term) d = 4 (this is the common difference between each successive number). So, the nth term is $5 + (n-1) \times 4 = 5 + 4n - 4 = 4n + 1$,

Other types of number patterns you should be familiar with

Square Numbers

1, 4, 9, 16, 25 ……..

Clearly this simply the square of natural numbers. The nth term is n^2

Cube numbers

Similarly 1, 8, 27, 64, …….is the cube of natural numbers. The nth term is n^3

Fibonacci Series

1, 1, 2, 3, 5, 8, 13, 21 …… (To get the next number, add the previous two) e.g. 1 +1 = 2, 1+2 =3, 2+3 = 5, and so on.

Multiply or divide each number by the same number

(1) 2, 6, 18, 54 ….. each number is 3× the previous number
(2) 12, 6, 3, 1.5 …each number the previous number divided by 2

Matrices

A matrix is simply an array of numbers

As we will see matrices can be very useful for geometrical transformations such as reflections, rotations and enlargements.

Example 1: $\begin{bmatrix} 2 & 1 \\ 3 & 2 \end{bmatrix}$ This is a 2 x 2 matrix. In other words it has two rows and 2 columns

Example 2: $\begin{bmatrix} 1 \\ 2 \end{bmatrix}$ This is a 2 × 1 matrix. It has two rows and one column

So you can see that matrices (plural of matrix) can come in different number of rows and columns. **We need to only worry about these two types of matrices for this exam**

Adding and Subtracting Matrices

You can add or subtract two matrices only if they have the same number of rows and columns.

Example 1: Add matrix **A** and matrix **B** where $A = \begin{bmatrix} 2 & 1 \\ 3 & 2 \end{bmatrix}$ and $B = \begin{bmatrix} 1 & 0 \\ 2 & 1 \end{bmatrix}$

Method: Simply add the elements of each matrix as shown below:

$$A + B = \begin{bmatrix} 2 & 1 \\ 3 & 4 \end{bmatrix} + \begin{bmatrix} 1 & 0 \\ 2 & 1 \end{bmatrix} = \begin{bmatrix} 3 & 1 \\ 5 & 5 \end{bmatrix}$$

Example 2: Now consider subtracting matrix B from matrix A

$$A - B = \begin{bmatrix} 2 & 1 \\ 3 & 4 \end{bmatrix} - \begin{bmatrix} 1 & 0 \\ 2 & 1 \end{bmatrix} = \begin{bmatrix} 1 & 1 \\ 1 & 3 \end{bmatrix}$$

Scalar Multiplication (Multiplying a matrix by a number) as shown below:

To multiply a matrix by a single number is very easy:

For example: $3 \times \begin{bmatrix} 1 & 3 \\ 2 & 1 \end{bmatrix} = \begin{bmatrix} 3 & 9 \\ 6 & 3 \end{bmatrix}$

These are calculated as shown:
$$3 \times 1 = 3 \quad 3 \times 3 = 9$$
$$3 \times 2 = 6 \quad 3 \times 1 = 3$$

We call the number ("3" in this case) a **scalar**, and this is called "scalar multiplication".

Multiplying a Matrix by another Matrix

Example 1: Multiply matrix $A = \begin{vmatrix} 1 & 3 \\ 2 & 1 \end{vmatrix}$ with matrix $B = \begin{vmatrix} 3 & 0 \\ 1 & 2 \end{vmatrix}$

Method: $\begin{vmatrix} 1 & 3 \\ 2 & 1 \end{vmatrix} \times \begin{vmatrix} 3 & 0 \\ 1 & 2 \end{vmatrix}$ to work this out you take each number from the first row of the first matrix and multiply it by each number of the first column in the second matrix in order to get the new first element/number of the new matrix shown. You repeat this process for each row and column respectively as shown below. **It is simpler than it sounds!**

$$\begin{vmatrix} 1 & 3 \\ 2 & 1 \end{vmatrix} \times \begin{vmatrix} 3 & 0 \\ 1 & 2 \end{vmatrix} = \begin{vmatrix} 1 \times 3 + 3 \times 1 & 1 \times 0 + 3 \times 2 \\ 2 \times 3 + 1 \times 1 & 2 \times 0 + 1 \times 2 \end{vmatrix} = \begin{vmatrix} 6 & 6 \\ 7 & 2 \end{vmatrix}$$

Multiply each number in the first row of A with each number in the first column of B and add the result as shown by the horizontal and vertical lines. Repeat this process for first row and the second column and so on. **The results are shown above**.

Example 2: Given that $\begin{vmatrix} 2 & a \\ 1 & b \end{vmatrix} \times \begin{vmatrix} 1 & c \\ 2 & -1 \end{vmatrix} = \begin{vmatrix} 8 & -3 \\ 3 & -1 \end{vmatrix}$. Find the values of a, b and c

Method: Multiply the two matrices together in the usual way that is in (i) we multiply each member of the first row in the first matrix by the first column in the second matrix and we get:

(i) $2 + 2a = 8 \implies 2a = 6 \implies a = 3$

(ii)   We carry on the multiplication process (1st row in the first matrix by 2nd column in the second matrix) to get

$2c - a = -3 \implies 2c - 3 = -3 \implies 2c = 6 \implies c = 3$

(iii)   Finally, multiplying 2nd row of the first matrix by the first column of the second matrix we get $\implies$ + 2b = 3 $\implies$ 2b = 2   b = 1

So we find that a = 3, b = 1 and c = 3.

Just to check substituting for a, b and c let us multiply the two matrices together: $\begin{vmatrix} 2 & 3 \\ 1 & 1 \end{vmatrix} \times \begin{vmatrix} 1 & 3 \\ 2 & -1 \end{vmatrix} = \begin{vmatrix} 2+6 & 6-3 \\ 1+2 & 3-1 \end{vmatrix} = \begin{vmatrix} 8 & 3 \\ 3 & -1 \end{vmatrix}$ which is in fact the result given in the example above.

Now let us look at Identity and zero matrices.

Identity Matrix: The identity matrix I for a 2×2 matrix is simply $\begin{vmatrix} 1 & 0 \\ 0 & 1 \end{vmatrix}$

The property of the identity matrix is that if you multiply a matrix A with the identity matrix I you get A. In other words A×I = A

Proof: If $A = \begin{vmatrix} a & b \\ c & d \end{vmatrix}$ and $I = \begin{vmatrix} 1 & 0 \\ 0 & 1 \end{vmatrix}$

Then $A \times I = \begin{vmatrix} a & b \\ c & d \end{vmatrix} \times \begin{vmatrix} 1 & 0 \\ 0 & 1 \end{vmatrix} = \begin{vmatrix} a+0 & 0+b \\ c+0 & 0+d \end{vmatrix} = \begin{vmatrix} a & b \\ c & d \end{vmatrix}$

Zero Matrix

This simply when the all the numbers inside a matrix are zero. So for a 2×2 matrix the zero matrix is $\begin{vmatrix} 0 & 0 \\ 0 & 0 \end{vmatrix}$. Yes you have guessed right if you multiply a 2×2 matrix with a 2×2 zero matrix you get a zero matrix

Matrices can also be used for transformations of images

When we want to create a reflection image we multiply the vertex matrix of our figure with what is called a reflection matrix. The most common reflection matrices are:

A reflection in the x-axis is achieved by multiplying the appropriate co-ordinates by the matrix below:

$$\begin{bmatrix} 1 & 0 \\ 0 & -1 \end{bmatrix}$$

Similarly a reflection in the y-axis is achieved by: $\begin{bmatrix} -1 & 0 \\ 0 & 1 \end{bmatrix}$

For a reflection in the origin we use: $\begin{bmatrix} -1 & 0 \\ 0 & -1 \end{bmatrix}$

Finally for a reflection in the line y=x, we use $\begin{bmatrix} 0 & 1 \\ 1 & 0 \end{bmatrix}$

<u>We can also rotate and enlarge shapes.</u>

<u>Example 1</u>: Show that P (1, 1) is rotated 90° clockwise about the origin by the matrix $\begin{bmatrix} 1 & 0 \\ 0 & -1 \end{bmatrix}$

Method: multiply $\begin{bmatrix} 1 & 0 \\ 0 & -1 \end{bmatrix}$ by $\begin{bmatrix} 1 \\ 1 \end{bmatrix} = \begin{bmatrix} 1 & 0 \\ 0 & -1 \end{bmatrix} \times \begin{bmatrix} 1 \\ 1 \end{bmatrix} = \begin{bmatrix} 1 \\ -1 \end{bmatrix}$

You can see in the graph below that the P(1, 1) has been rotated clockwise by 90° to Q(1,-1)

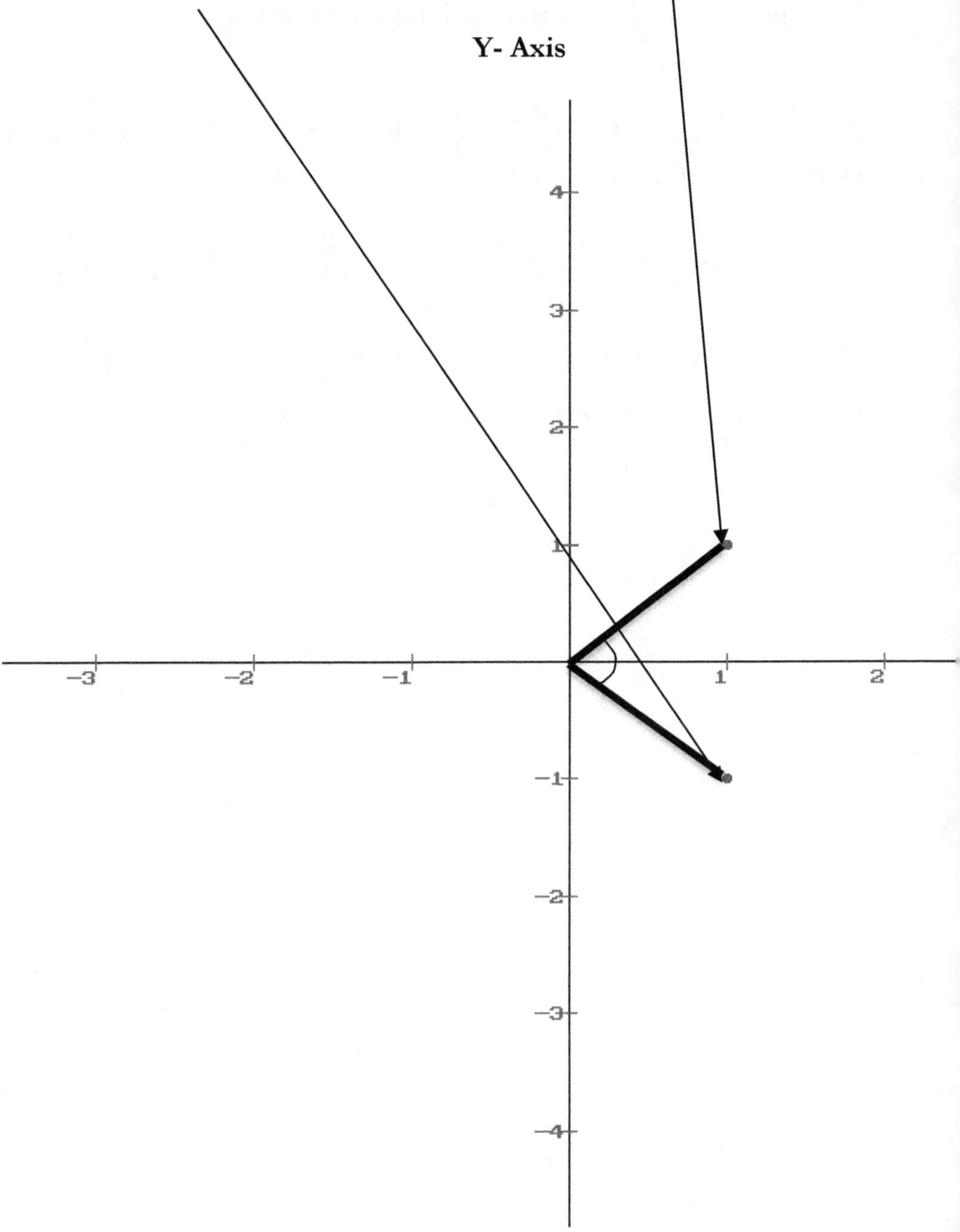

Example 2: Transforming co-ordinates by two consecutive matrices.

$P = \begin{bmatrix} 2 & 0 \\ 0 & 2 \end{bmatrix}$ and $Q = \begin{bmatrix} -1 & 0 \\ 0 & 1 \end{bmatrix}$ The point A (1, 2) is transformed by the matrix QP to A'. Find the resulting transformation A'.

Method: First work out QP. $\begin{bmatrix} -1 & 0 \\ 0 & 1 \end{bmatrix} \times \begin{bmatrix} 2 & 0 \\ 0 & 2 \end{bmatrix} = \begin{bmatrix} -2 & 0 \\ 0 & 2 \end{bmatrix}$

Now multiply this by $A = \begin{bmatrix} 1 \\ 2 \end{bmatrix}$. So we get $A' = \begin{bmatrix} -2 & 0 \\ 0 & 2 \end{bmatrix} \times \begin{bmatrix} 1 \\ 2 \end{bmatrix} = \begin{bmatrix} -2 \\ 4 \end{bmatrix}$. So the resulting transformation A' = (-2, 4)

Practice Questions on Matrices

(1) If $A = \begin{bmatrix} 2 & 3 \\ -1 & 2 \end{bmatrix}$ and $B = \begin{bmatrix} 4 & -1 \\ 2 & 3 \end{bmatrix}$

Work out:

(a) AB

(b) BA

(c) 3B

(d) A^2

(2) Given that $\begin{bmatrix} 2 & x \\ -1 & 3 \end{bmatrix}\begin{bmatrix} 2 \\ 6 \end{bmatrix} = \begin{bmatrix} 12 \\ 16 \end{bmatrix}$ work out the value of x

(3) If $\begin{bmatrix} 2 & x \\ 3 & 1 \end{bmatrix}\begin{bmatrix} 1 & 2 \\ 3 & y \end{bmatrix} = \begin{bmatrix} 11 & 16 \\ z & 10 \end{bmatrix}$ Find the values of x, y and z

(4) If $M = \begin{bmatrix} 3 & 4 \\ -1 & 2 \end{bmatrix}$ show that IM = M where I is the identity matrix

(5) The co-ordinates x, y are transformed by the matrix $\begin{bmatrix} 0 & -1 \\ -1 & 0 \end{bmatrix}$ to give x', y'. What are the values of x', y' and what type of transformation is this?

(6) B(x,y) is transformed to the point B'(-1, 0) by the matrix $\begin{bmatrix} 2 & 4 \\ 1 & 1 \end{bmatrix}$. Work out the values of x and y.

Answers to Practice Questions on Matrices

1 (a) $\begin{bmatrix} 14 & 7 \\ 0 & 7 \end{bmatrix}$

(b) $\begin{bmatrix} 5 & 10 \\ 1 & 12 \end{bmatrix}$

(c) $\begin{bmatrix} 12 & -3 \\ 6 & 9 \end{bmatrix}$

(d) $\begin{bmatrix} 1 & 12 \\ 0 & 1 \end{bmatrix}$

2   x = 1

3   x = 3, y = 4 and z = 6

4   M×I = $\begin{bmatrix} 3 & 4 \\ -1 & 2 \end{bmatrix} \times \begin{bmatrix} 1 & 0 \\ 0 & 1 \end{bmatrix} = \begin{bmatrix} 3+0 & 0+4 \\ -1+0 & 0+2 \end{bmatrix} = \begin{bmatrix} 3 & 4 \\ -1 & 2 \end{bmatrix}$

5   $x = \frac{1}{2}$ and $y = -\frac{1}{2}$

6   The resulting transformed co-ordinates are x = 28 and y = 8

Statistics

Sampling

In order to carry out a statistical survey or investigation we obviously need to collect data.

In an ideal situation you would survey the **whole population** that is of interest, but in reality you can only survey a proportion of the population. This is called a **sample** of the **population.**

The most important thing to remember is that the **sample is representative and not biased**. This is best achieved by taking a random sample. (A random sample simply means every member has an equal chance of being included. Finally, the sample has to be reasonably large compared to the size of the population.)

Sampling methods:

Random Sampling: This simply means pick a sufficient number of members randomly from the group in question. (In principle you can assign a number to every individual in the population, generate random numbers say by using a random number generator in a spreadsheet or scientific calculator and then match the random numbers to the members of the population.)

Stratified Sampling: When a population can be split into groups for example by educational attainment or income levels you can use Stratified Sampling.

Example: In a given Primary school the number of boys and girls in each year group is shown below:

	Year 5	Year 6
Boys	56	44
Girls	54	46

A sample of 40 pupils is taken stratified by the appropriate year grop and gender. Calculate the number of year 5 boys that should be in the sample.

First let us find the proportion of boys that are in year 5.

The total population (in this case the number of pupils) = 56 + 44 + 54 + 46 = 200

Now we need to find the proportion of pupils that are year 5 boys. This is $\frac{56}{200} = \frac{28}{100} = 0.28$

Since the total sample size is 40 the number of year 5 boys that need to be included is 0.28×40 =2.8×4 = 11.2 or **11 rounded to the nearest whole number.**

Surveys and Questionnaires

Although you can have closed questions or open questions these questions must be sensibly framed.

Example 1: What is your age? What is your income? These are obviously open questions

Example 2: Do you watch TV? (Yes or No) Do you play football? (Yes or No) These are closed questions

Example 3: A teacher is trying to find out how often pupils were set homework the previous week in Year 9 and how many of them completed doing it in that week The questionnaire is given to 24 pupils in one class of year 9. The whole year 9 group consists of 93 pupils. How could you improve on selecting the pupils as well as the survey questionnaire below?

The questionnaire is as follows:

(1) Was homework set regularly?
(2) Are you set homework in all subjects?
(3) Did you complete them?

Clearly a few improvements could be made:

(A) Firstly the pupils in all year 9 group could be assigned numbers, random numbers could be generated and matched to the 93 pupils and 24 pupils picked. **This would make it a less biased sample.**
(B) Secondly the questionnaire could be improved as shown below:

(a) How many times were you set home works <u>last week</u>: (a) None (b) 1 – 2 times (c) 3 – 4 times (d) more than 4 times

(b) How many home works did you finish?
(a) None (b) 1 - 2 (c) 3 - 4 (d) All of them

Notice that in the second questionnaire you are given a specific time frame and a reasonable range of options. So the teacher can collect more accurate data just for that week!

Example 4: What is wrong with the same question phrased slightly differently?

(1) How many times were you set homework last week?
(a) None (b) 1 – 2 (c) 2 – 3 (d) more than 4
(2) How many did you finish?
(a) None (b) 1 - 2 (c) 2 -3 (d) more than 4

Notice the ranges given in (b) and (c) are inclusive. If someone was set or pupils did 2 homework they could answer either (b) or (c). This would cause confusion in analysing the data.

<u>**Things to remember when sampling:**</u>

(1) Make it as unbiased as possible
(2) Make sure there is a time frame (e.g. how many times did you buy ice cream <u>last month,</u> <u>last week</u> or some other sensible time frame. Rather than how many times did you buy ice cream?
(3) Make sure the questionnaire has no overlapping data when using class intervals
(4) Try to make the question as clear as possible
(5) Be careful with open ended questions that are difficult to analyse

Averages

First consider the different type of 'averages'

As Mean, Median, Mode and Range

Mean: The sum of the numbers in a data set divided by the number of numbers in the set

Median: The middle number of a data set when listed in order

Mode: The most frequently occurring number or numbers in a data set

Range: The difference between the highest and the smallest numbers in a data set

Note: For a frequency distribution the Mean $= \dfrac{\sum fx}{N}$

(See Example7)

Example1:

Find the mean value of the following data set:

2, 7, 1, 1, 7, 8, 9

Method: Find the sum first

$2 + 7 + 1 + 1 + 7 + 8 + 9 = 35$

Now divide this total by 7, since this is the total number of numbers

So, $35/7 = 5$

Hence the mean value of this data set is 5

Example 2:

Find the median of 3, 7, 1, 8 and 6

Method: First re-order from smallest to biggest

Re-writing the numbers we have: 1, 3, 6, 7, 8

Clearly the middle number is 6.

Hence, the median is 6

Example 3:

Find the median of 3, 6, 7, 1, 8 and 5

First re-arrange to get 1, 3, 5, 6, 7, 8

In this case the middle number is between 5 & 6

So the median is $(5 + 6)/2 = 5.5$

Example 4:

Find the Range of the data set 3, 5, 7, 1, 8, and 11

Method: find the difference between the biggest and smallest numbers

So the Range = $11 - 1 = 10$

Example 5:

Find the Mode of the following numbers:

1, 4, 4, 4, 7, 8, 9, 9, 11, 12

The most frequently occurring number is 4

Hence the Mode is 4

Example 6:

Find the mode of 1, 3, 3, 3, 3 5, 5, 5, 5, 8, 8, 9

Clearly there are two modes here:

Both '3' and '5' occur most frequently

So we say this is a bi-modal distribution.

That is, a distribution with two modes, namely 3 and 5

Example 7

Frequency table:

Find the mean of the following frequency distribution

x (value)	1	2	4	6
f (frequency)	4	3	5	1

Finding the mean:

Step1: Multiply the frequency (f) by the value (x) to find fx

fx = 1×4 + 2×3 + 4×5 + 6×1 = 4 + 6 + 20 + 6

(Note: we obtained the above results from the frequency table, e.g. 4x1=4, 2X3=6, 4X5=20 and 6X1=6)

Step2: Add these results to find $\sum$ fx

$\sum$ fx = 4 + 6 + 20 +6 = 36

Step 3: Add the frequencies to find the total frequency N

$$N = \sum f = 4 + 3 + 5 + 1 = 13$$

Step 4: Divide $\sum fx$ by N to obtain the mean

$$\bar{x} = (\text{Mean}) = \frac{\sum fx}{N}$$

So the mean value is 36/13 = 2.77

Stem and Leaf Diagram

Example: A biology test is marked out of 50. The marks that the pupils get are as follows: 8, 32, 43, 37, 20, 22, 24, 10, 18, 30, 11, 27, 5. Draw a stem and leaf diagram and work out the median.

The stem and leaf diagram is shown below:

```
0 | 5  8
1 | 0  1  8
2 | 0  2  4  7
3 | 0  2  7
4 | 3
```

Key: 3 | 7 means 37

The median is the middle number which is 22.

Grouped Frequency

Example

60 School Children were weighed. The data is shown in the grouped frequency table below. Find the estimated mean value of the weight correct to 2 decimal places.

Weight (w Kg)	Frequency (f)	Mid-point(w)	f×mid-point
$30 < w \leq 40$	8	35	8×35= 280
$40 < w \leq 50$	16	45	16×45=720
$50 < w \leq 60$	18	55	18×55=990
$60 < w \leq 70$	12	65	12×65=780
$70 < w \leq 80$	6	75	6×75=450
	60		= 3220

Method: To find the mean sum up frequency×mid-point and divide by the total frequency. So in this case the mean = 3220÷60 =

322÷6 =161÷3 = 53.67.   **Hence, the mean weight = 53.67kg**

Scatter Graphs

Scatter graphs or diagrams are used to show the type of relationship between two variables, for example height and weight, maths and physics scores, reading scores and IQ and so on. It also gives information on the type of correlation between the two variables. The 'line of best fit' goes through close to most of the points. We can work out the equation of this 'line of best fit' as we shall see later.

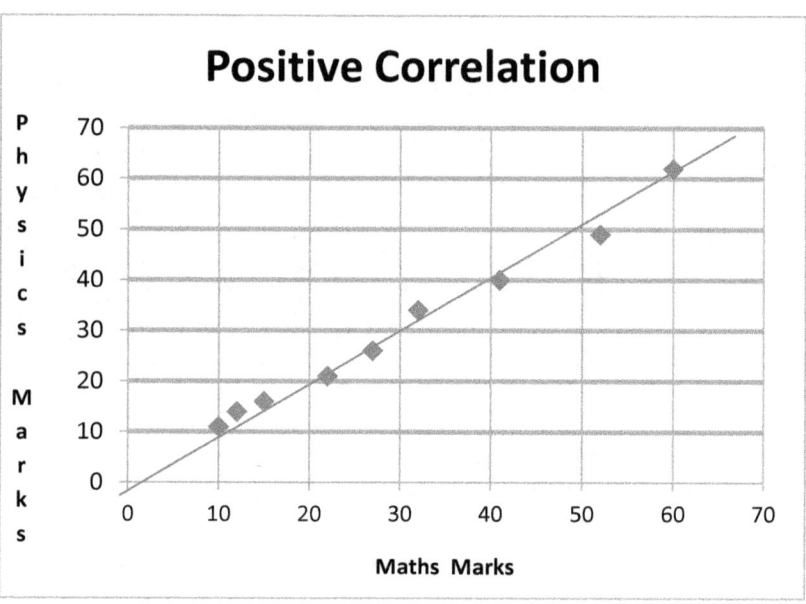

Positive Correlation

This means when the value of one variable increases so does the value of the other one as shown in the example above. A certain number of pupils' test marks in maths and physics are plotted. (The straight line drawn is called the line of best fit.) . You can see in this case there is a positive correlation between Maths marks and Physics marks.

> **(1) How many pupils get more than 45 marks both in Maths and Physics?**
>
> Answer: 2 pupils get above 45 marks in both these subjects.

Method: look at the horizontal and vertical axis and draw an imaginary line at 45. Above 45 marks in both subjects you can see the record of two pupils.

(2) How many pupils get less than 20 marks in both Maths and Physics?

Answer: 3 pupils get less than 20 marks (using similar reasoning to the first answer)

Negative Correlation

This means that as one variable decreases in value the other one increases.

Example of Negative correlation:

This time some pupils' scores in Maths were plotted with their scores in History. It appears that in this particular case there was a negative correlation between Maths and History scores as shown below.

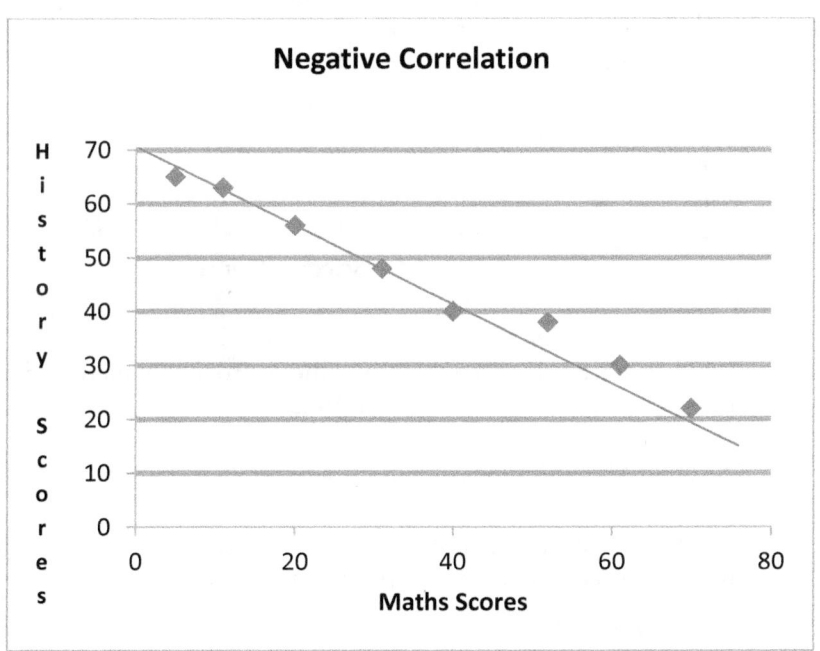

Another example of a negative correlation is the one between the price of a car and its age. As a car gets older its price is generally lower.

Zero Correlation

This is when there is no relationship between the two variables. The points are scattered all over the place so that we cannot really draw a line of best fit. For example consider the relationship between Science and English marks shown below in this particular example

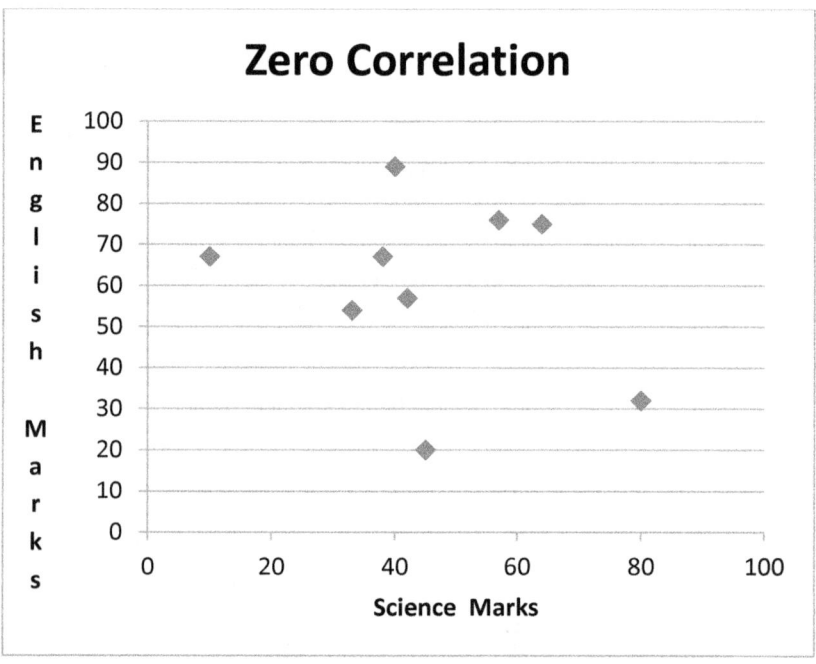

(1) In the example above what was the approximate mark in Science for the pupil who scored 20 marks in English?

Answer: 45 is the approximate mark in Science

Method: From the vertical axis, go along the horizontal line at 20 marks, this corresponds to approximately 45 marks in Science

Finding the equation of the 'line of best fit'

Consider two variables that are positively correlated for example maths and physics marks. 10 students take both these tests and the marks are shown in the table below. The line of best fit goes through the first (**12, 10**) and the last point (**28, 30**) in the respective marks as shown in the table below.

Maths marks (x-axis)	**12**	13	14	16	20	25	15	28	30	**28**
Physics marks (y- axis)	**10**	12	15	18	19	24	16	26	31	**30**

The equation of a straight line is given by y - a = m(x –b), where 'a' and 'b' are point it goes through and m is the gradient.

Let us find the gradient first, taking the last and first point. $M = \frac{30-10}{28-12} = \frac{20}{16} = \frac{5}{4}$

So we can now write the equation as $y - a = \frac{5}{4}(x - b)$

Let's take one point it goes through, e.g. (12, 10). Hence the line of best fit is approximately $y - 10 = \frac{5}{4}(x - 12)$. This simplifies to $y - 10 = \frac{5}{4}x - \frac{60}{4}$, or $y - 10 = \frac{5}{4}x - 15$ and finally $y = \frac{5}{4}x - 5$. How does this equation help?

If we know the maths marks of student is 24, we can estimate the marks in physics this student is likely to have got. Substitute 24 for x in the equation $y = \frac{5}{4}x - 5$ this means $y = \frac{5}{4} \times 24 - 5$ so y = 30 - 5 = 25. This means the student is expected to get around 25 marks in Physics.

Note: If you had chosen (28, 30) as your x and y co-ordinates, although the answer would be a little different the method is still correct and you would still get full marks. Remember this is simply called the **'line of best fit'** and is an **estimate**.

Note: If you are **not told which point(s)** the line of best fit goes through you can simply draw a line that goes through most of points in the scatter graph (If possible half the points on the scatter graph should be above the line of best fit and the other half should be below the line of best fit. **This method gives you a rough line of best fit and is quite acceptable for your exam question at GCSE level.**

Box and Whisker plots

A box and whisker plot summarizes the **spread** of data. It shows the **median**, the **upper** and **lower quartile** as well as the **lowest** and **highest** value of the data set.

Remember the median is the middle value of the data set – this means half the data set is below and the other half above. The lower quartile is at the 25% of the data set. Similarly, the upper quartile is at 75% of the data set. The difference between the upper and lower quartile is called the Inter Quartile Range (IQR).

The lower quartile is often referred to as Q1 and the Upper quartile as Q3

The interquartile range is simply Q3 – Q1 that is Upper Quartile – Lower Quartile

However note that the range is simply the difference between the highest and lowest number. Consider some examples below:

Example 1:

The box and whisker diagram below represents the following marks in Science for 10 pupils. The marks are out of 40.

Pupil Marks In Science	14	22	12	16	34	6	28	30	8	5

(1) Let's find the median first. Arrange the marks in order so we get 5, 6, 8, 12, 14, 16, 22, 28, 30, and 34

The median is the middle number so it is between 14 and 16. (14+16)/2 = 30/2 =15

The upper quartile is three quarters of the way up (above this is the highest 25%). Another way of looking at this is that the upper quartile is the median value of the top half that is 16, 22, 28, 30 and 34. Which means the upper quartile is the middle number that is 28. Similarly the bottom half numbers are:

5, 6, 8, 12, 14 which means the Lower Quartile is the middle number of this data set which is 8.

The Inter Quartile Range (IQR) = Upper Quartile – Lower Quartile = 28 – 8 = 20

The range is however defined as the highest score minus the lowest score, so the range in this case is 34 -5 =29

The box and whiskers diagram below illustrates this data as a visual representation.

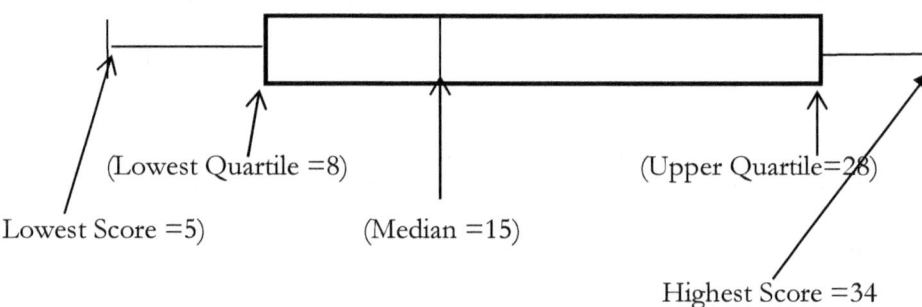

(Note the **ends of the Whiskers** denote the **lowest** and the **highest** scores. **The ends** of the **rectangle** represent the **lowest** and the **highest quartiles** as shown. Finally, the median is the **vertical line** inside the rectangle where it shows median =15)

Cumulative Frequency diagrams

Example 1

A class test in maths was marked out 70. The table below shows the distribution of marks among 20 pupils.

Marks (Max 70 marks)	Number of Pupils (frequency)	Cumulative Frequency (keep adding the frequencies)
1 - 10	0	0
11 - 20	2	0 + 2 = 2
21 - 30	3	2 + 3 = 5
31 - 40	5	5 + 5 = 10
41 - 50	5	10 + 5 = 15
51 - 60	4	15 + 4 = 19
61 - 70	1	19 + 1 = 20

[Grab your reader's attention with a great quote from the document or use this space to emphasize a key point. To place this text box anywhere on the page, just drag it.]

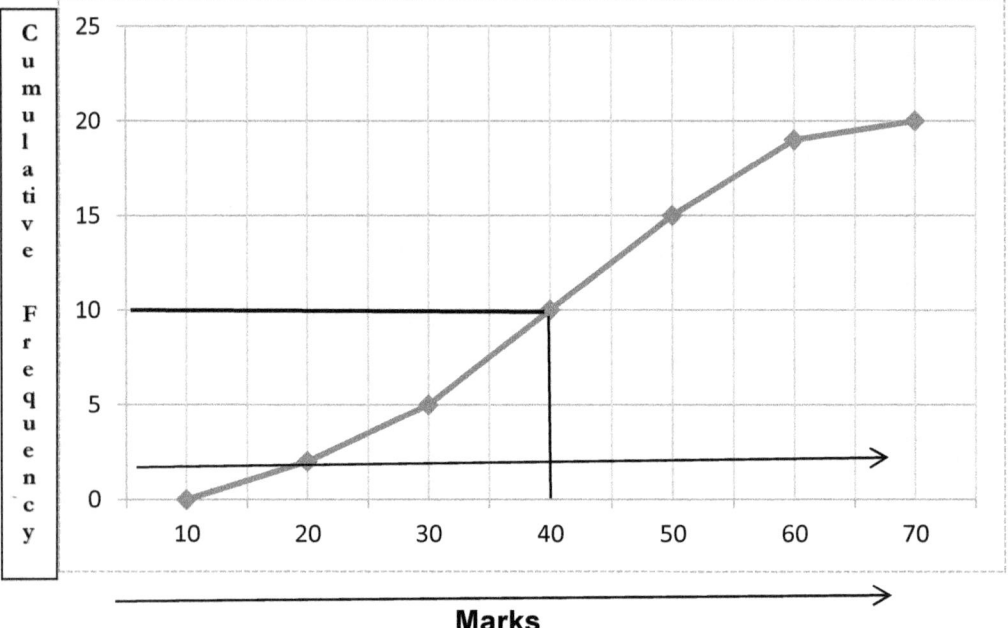

From the graph you can see the median (half way up the cumulative frequency is 10) this corresponds to 40 marks. This means half the students get up to 40 marks and half get over 40 marks.

(Just for your information when plotting the graph **one plots the upper bound of the mark with the corresponding cumulative frequency value.** So for example in the mark range 21- 30 you plot 30 (the upper bound) with 5)

Now, see if you can follow the answers to the questions below

(1) What mark corresponds to the lower quartile?

The lower quartile is a quarter of the way up the cumulative frequency axis. This corresponds to 5. The total cumulative frequency is 20, so a quarter on this axis is at 5. If you draw a horizontal line at 5 until it meets the curve and then draw the corresponding vertical line at this point it meets the horizontal line at 30. This means 25% of the pupils score 30 marks or below.

(2) What marks correspond to the upper quartile?

The upper quartile is three quarters of the way up the cumulative frequency axis. Three quarters of 20 (the total cumulative frequency) is 15. This corresponds to 50 marks on the horizontal axis.

This means that 25% of the pupils get more than 50 marks.

(3) What is the Inter Quartile Range (IQR)?

This is simply the difference between the Upper Quartile and the Lower Quartile.

IQR = Upper Quartile – Lower Quartile

IQR = 50 - 30 = 20

Summary

To find Median: From the vertical axis (which represents the cumulative frequency) go up to 50% or half way up this axis and draw a horizontal line to the cumulative frequency curve, then draw a vertical line at this point to meet the horizontal axis and read of the appropriate value

To find the Upper Quartile go up the 75% mark vertically (three quarters of the way up) and similarly read the corresponding value on the horizontal axis.

To find the Lower Quartile go up the 25% mark vertically (one quarter of the way up) and read the corresponding value on the horizontal axis

To find the Inter Quartile Range simply take the difference the Upper Quartile and the Lower Quartile.

Histograms

These are similar looking to bar charts but have some basic differences

(1) You need to remember that there are **no gaps between the bars**, as the data shown is continuous
(2) The bars can be either **equal widths or different widths** depending on the class interval
(3) The vertical axis can be used to estimate frequencies only when the class intervals are equal otherwise you need to work out the frequency density

Example: The following marks were obtained on a maths class test as shown by the table below. Show the information as a histogram.

(Note: In this case the class intervals have the <u>same width</u>)

Marks	$10 \leq m < 20$	$20 \leq m < 30$	$30 \leq m < 40$	$40 \leq m < 50$
Frequency	4	6	14	2

Reading the marks in the table:

The class interval $10 \leq m < 20$ means the marks are between 10 and 20 (more precisely, the table shows that 4 pupils got greater than or equal to 10 marks and less than 20 marks.

A typical histogram with the **same class interval** is shown below:

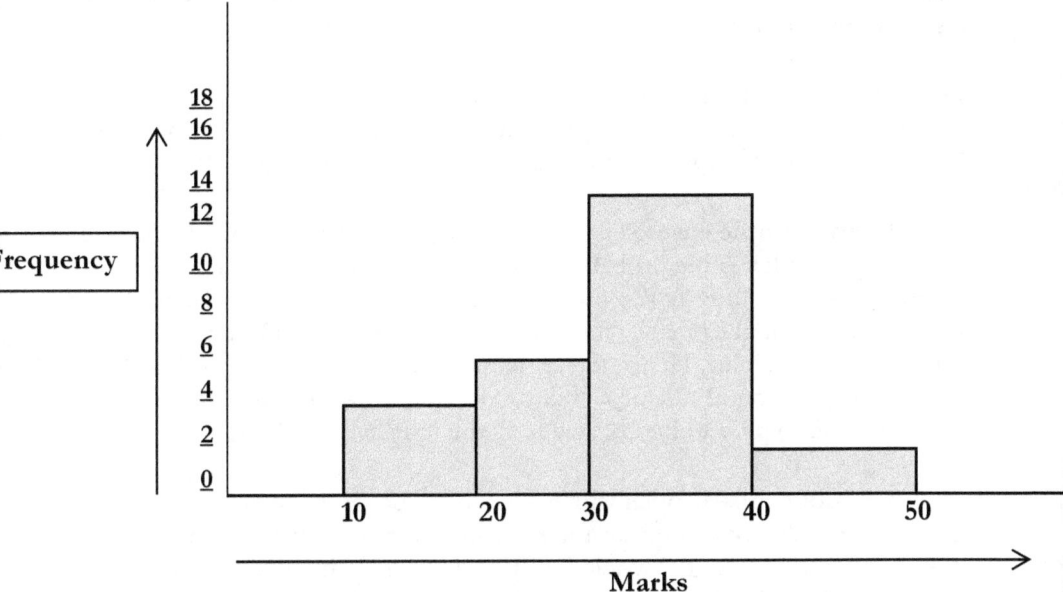

However, the class intervals (the widths) can also vary.

For histograms with **unequal class widths** it is important to note that the **frequency is represented by the area of the bars** and **not the height**.

In order to work out the height of a bar you have to calculate the frequency density of each data class using the formula: **Frequency density = Frequency ÷ Class width**

However, if the class intervals were different like below:

Marks	$10 \leq m < 20$	$20 \leq m < 40$	$40 \leq m < 50$	$50 \leq m < 70$
Frequency	4	6	14	2
Frequency density (freq÷class width)	0.4 (4÷10) = 0.4	0.3 (6÷20) = 0.3	1.4 (14÷10) = 1.4	0.1 (2÷20) = 0.1

You can work out the frequency density (height) as shown above.

Probability

Probability is defined as the likelihood of an event happening. Probability lies between 0 and 1.

A probability of 0 means that an event will definitely not happen or it is impossible to happen. Likewise a probability of 1 means is certain to happen. Probability is usually expressed as a fraction, a decimal or a percentage.

Consider two simple cases: There are 4 blue balls in a bag. You take out a ball at random. (1) What is the probability that the ball you pick is red? (2) What is the probability that the ball you pick is blue? Although this is a trivial example you can see that in question (1) it is impossible to pick red ball since all the 4 balls in the bag are blue. Hence the probability of picking up a red ball is 0. Similarly, in question (2) the probability that you pick a blue ball is 1. That is you are certain to pick a blue ball, since all the four balls are blue.

Many events of course happen with a probability between 0 & 1. For example a probability of 0.8 would indicate a fairly high chance of an event happening, whereas a probability of 0.2 would imply a low probability of an event happening. The probability of an event happening is defined as:

$$\frac{number\ of\ ways\ in\ which\ the\ event\ can\ happen}{total\ number\ of\ outcomes}$$

Also note that the probability of an event **not happening** is **1 – the probability of an event happening**

Notation used: P(A) means probability of event A happening. Hence probability of event A not happening would be 1 – P(A).

Typical examples:

Example 1:

There are 5 red, 6 green and 7 blue beads in a bag.

(1) You pick a bead at random from the bag. What is the probability of picking a red bead? Answer $P(R) = \frac{5}{18}$ (Reason: there are 18 beads altogether, and 5

of them are red, so the chance or probability of picking a red bead is 5 in 18 or $\frac{5}{18}$)

(2) What is the probability of picking a green or blue bead? Answer $P(G \text{ or } B) = \frac{13}{18}$

Reason: there are 18 beads altogether, and the number of green and blue beads combined total 13. Hence the probability of picking a green or blue bead is 13 in 18 or $\frac{13}{18}$.

(3) What is the probability of not picking a green bead? Answer: $P(\text{not } G) = 1 - P(G) = 1 - \frac{6}{18} = \frac{12}{18}$

A simpler way of doing the same problem is to say that since there are 18 beads altogether and 6 of them are green, then this means that 12 are not green, hence the probability of not picking up a green bead is 12 in 18 that is $\frac{12}{18}$. You could of course simplify $\frac{12}{18}$ to $\frac{2}{3}$ (dividing both the top number 12 and bottom number 18, by 6)

Relative Frequencies:

When you do not know the probability exactly you can use an experimental method of relative frequencies to assess an **estimate** of the probability. Let's say you are not sure whether a die is fair or biased. You test it out by throwing it 200 times and get the number 6, fifty times.

Relative frequency of an event is defined as

$$= \frac{\text{Number of times the event happened}}{\text{Total number of trials}}$$

In this example, using the formula above the relative frequency $= \frac{50}{200} = \frac{5}{20} = \frac{1}{4}$ = 0.25, it seems that the die is biased as we would expect the number 6 to occur roughly $\frac{1}{6} \times 200$ times which is around 33 times! (Just for interest to be absolutely sure that the result wasn't just a coincidence we may need to repeat the experiment again with a bigger sample and also do something called 'significance' testing to make sure that the die is really biased)

Expected Number

Example 1: If a fair die is thrown 660 times approximately how many threes are we likely to get?

The probability of getting any number when a fair die is thrown is $\frac{1}{6}$. The expected number of threes = P(3) × 660 = $\frac{1}{6}$ x 660 = 110. We would therefore expect the number three to occur 110 times.

Example 2: The probability of passing the QTS numerical skills test the first time is 0.85. If 1000 teacher trainees take this test in a certain region over a year, how many do you expect to pass this test the first time?

Method: 1000 × 0.85 = 850. So we would expect 850 teacher trainees to pass the QTS test the first time in this particular region.

Multiplication law in probability

When you have independent events (that is the outcome of one is not affected by the outcome of the other) then to find the probability of say event A and event B happening we simply multiply the probabilities of A and B together.

Example 1: What is the probability that we will get two sixes when a die is rolled two times?

Method: Probability that we get '6' followed by '6' = $\frac{1}{6} \times \frac{1}{6} = \frac{1}{36}$

Example 2: A fair coin is flipped three times. What is the probability it will turn up 'heads' on all three occasions? Give your answer to 2 decimal places.

Method: Probability that it turns up 'heads' **and** 'heads' **and** 'heads' = $\frac{1}{2} \times \frac{1}{2} \times \frac{1}{2} = \frac{1}{8}$ =0.125 or 0.13 to 2 decimal places. Or you could have calculated it another way i.e. 0.5×0.5×0.5 = 0.25×0.5 =0.125 and then give your answer to two decimal places 0.13 as required.

Example 3:

If a fair die is thrown twice what is the probability of getting a 'six' followed by 'not a six'. Give your answer as a fraction.

Method: P(getting a six) $= \frac{1}{6}$, so the probability of 'not getting a six' $= 1 - \frac{1}{6} = \frac{5}{6}$. Hence the probability that you get a 'six' followed by 'not a six' $= \frac{1}{6} \times \frac{5}{6} = \frac{5}{36}$

Addition law in probability: When two or more events are mutually exclusive (i.e. they cannot occur together), then the probability of A **or** B **or** C happening is simply found by adding the respective probabilities. That is p(A) + p(B) + p(C).

Example 1: there are 6 blue beads, 8 green beads and 15 black beads in a bag. What is the probability of picking either a green or a black bead?

Method: altogether there are 29 beads. Probability of picking a green bead $= \frac{8}{29}$, similarly the probability of picking a black bead $= \frac{15}{29}$. Hence the probability of picking either a green or black bead is found by adding the two probabilities together. P(Green or Black) $= \frac{8}{29} + \frac{15}{29} = \frac{23}{29}$

Example 2: A pack of 52 cards is shuffled. What is the probability of picking an ace or a king or a Queen? Give your answer in its simplest form?

Method: There are 52 cards altogether. There are 4 aces, 4 kings and 4 Queens in a pack of cards. Hence the probability of picking an Ace 'or' a King 'or' a Queen $= \frac{4}{52} + \frac{4}{52} + \frac{4}{52} = \frac{12}{52} = \frac{6}{26} = \frac{3}{13}$

Problems 'with' and 'without' replacement

Example 1: There are 3 red marbles and 7 blue marbles in a bag. A marble is picked at random. It is put back in the bag and another marble is picked at random. What is the probability of picking two red marbles consecutively?

You can see that this is a fairly straight forward problem. There are 10 marbles altogether and 3 of them are red. The probability of picking a red marble the first time is thus $\frac{3}{10}$. Since the marble that you pick the first time is now put back you have the same situation as before, so the probability of picking a red marble the second time is also $\frac{3}{10}$. Hence, the probability of picking a red marble followed by another red marble is $\frac{3}{10} \times \frac{3}{10} = \frac{9}{100}$

Example 2: Now consider this, let's say there are still 3 red marbles and 7 blue marbles in a bag. You pick up one but don't put it back in the bag. If you pick up a red marble the first time, then there are now only 2 red marbles and 7 blue marbles left. The probability of picking up a red marble the first time is still $\frac{3}{10}$ but the probability of picking up a red marble the second time is now $\frac{2}{9}$. (since there are now 9 marbles in total of which 2 are red). Hence the probability of picking up 2 red balls consecutively on this occasion is $\frac{3}{10} \times \frac{2}{9} = \frac{6}{90}$ or $\frac{1}{15}$ writing the fraction in its simplest form!

Tree diagrams are another way of solving probability problems showing two or more events.

Example: In year7 the probability that a pupil does his homework is $\frac{3}{8}$ and the probability that a pupil comes late to class is $\frac{5}{6}$. These two events are independent. Draw a tree diagram and find the probability that a given pupil does the homework and is late for class.

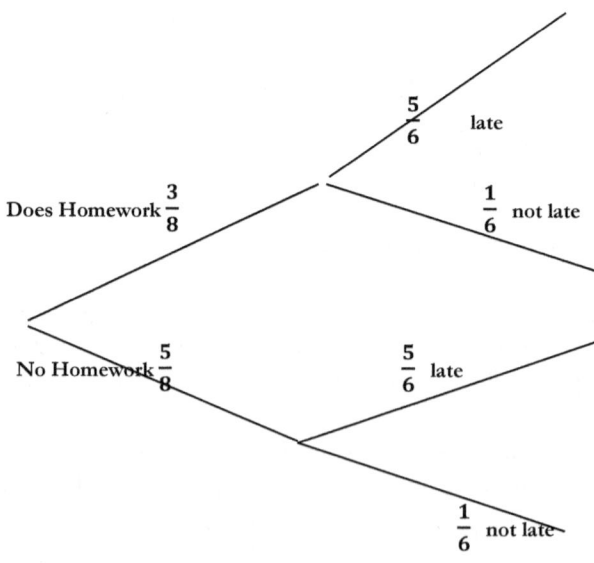

Method: To do this using the tree diagram we simply follow the branch 'does homework' **and** 'late' As you can see the probability in this case is $\frac{3}{8} \times \frac{5}{6} = \frac{15}{48} = \frac{5}{16}$ in its simplest form

Sample space diagrams

This is another way of listing the outcomes of two events

Example: Two dice are thrown together and their total scores are noted as shown below. From the sample space diagram (1) Find the probability that the score is 11 and (2) Find the probability that the score is less than 6.

		\multicolumn{6}{c}{First die}					
		1	2	3	4	5	6
	1	2	3	4	5	6	7
	2	3	4	5	6	7	8
Second	3	4	5	6	7	8	9
Die	4	5	6	7	8	9	10
	5	6	7	8	9	10	11
	6	7	8	9	10	11	12

Answers (1): $P(11) = \frac{2}{36} = \frac{1}{18}$

Method: If you look at the sample space diagram you can see that the possible outcomes are recorded as totals. There are two possible outcomes with a total score of 11 out all the 36 possible outcomes. Hence the probability of getting a total of 11 is $= \frac{2}{36}$ which simplifies to $\frac{1}{18}$.

Answer (2): $P(\text{less than 6}) = \frac{10}{36} = \frac{5}{18}$

Method: Again from the sample space diagram above you can see that all the totals less than 6 are 2, 3, 4 & 5 in the first line, 3, 4 & 5 in the second line, 4 & 5 in the third line and 5 in the fourth line. Hence there are 10 possible outcomes whose total scores are less than 6. The total number of possible outcomes is 36. Hence, the probability of getting a score of less than 6 is $\frac{10}{36}$ which simplifies to $\frac{5}{18}$.

Note: Just for interest P(less than 6) can also be written as P(<6) since the sign < means less than.

Summary:

(1) Probability lies between 0 and 1 and is usually expressed as a decimal, a fraction or a percentage. The probability of an event can never exceed 1.

(2) When events are independent, to find the probability of A <u>and</u> B occurring together we <u>multiply</u> the probabilities of the respective events. Remember the word <u>'and' is associated with '×' or multiplication</u>.

(3) When events are mutually exclusive the probability of A <u>or</u> B <u>or</u> C happening is found by <u>adding</u> the individual probabilities. Remember the word <u>'or' is associated with '+' or addition</u>.

(4) When working out probabilities consider whether it is 'with' or 'without' replacement

(5) You can generate tree diagrams or sample space diagrams to visualize probabilities and outcomes if it helps you.

Density Mass and Volume

Example:

Work out the density of a 12 kg piece of metal with a volume of 2.5m^3

Method:

Density = Mass ÷ Volume, so:

12 ÷ 2.5 = 4.8 kg per m^3.

You need to remember these formulas:

Density = Mass ÷ Volume

Mass = Density × Volume

Volume = Mass ÷ Density

You can also use the triangle below to help you remember the formulas:

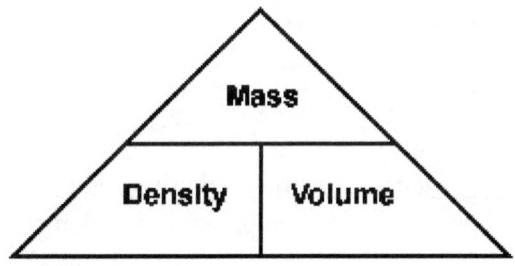

Practice questions 5 – no calculators allowed

(1) Find the mode, median and range of the following numbers:
7, 3, 1, 2, 2, 10, 15   (3 marks)

(2) Find the area of the triangle ABC shown. The base is 8cm and the height is 12 cm   (2 marks)

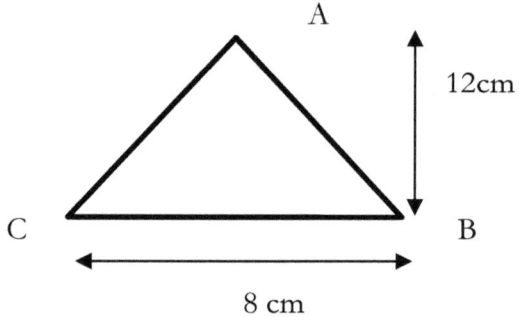

(3) Complete the following sentences for the circle theorem:
   (i) The angle at the centre = ... angle subtended at the circumference   (1 mark)
   (ii) The radius meets a circle tangent at ......   (1 mark)
   (iii) The angle between the tangent and the chord = the angle in the ....   (1 mark)
   (iv) The opposite angles of a cyclic quadrilateral add up to.......   (1 mark)
   (v) The triangle in a semi- circle has an angle of ...... at the circumference   (1 mark)

(4) Simplify the surd $\frac{1}{\sqrt{3}}$ (2 marks)

(5) The points P(1, 1) and Q(2, 3) lie on a straight line. Find (a) the mid-point of PQ and (b) find the equation of the line PQ

(4 marks)

(6) There are 5 blue beads, 7 green beads and 8 black beads in a bag. What is the probability of picking either a green or a black bead? (2 marks)

(7) A fair die is thrown 780 times what is the expected number of times it will land a 'six'? (2 marks)

(8) Find the exterior angle of a regular 20 sided polygon (3 marks)

(9) Two dice are thrown together and their total score is noted Find the probability that the total score is 10. (3 marks)

(10) A teacher represents the relationship between marks in a maths test and a science test by the scatter graph shown below. The Maths marks are out of 60 and the Science marks are out of 10. (a) What is the correlation between this particular maths and science test? (b) Draw a 'line of best fit' (c) When the pupil's mark in Science was 6, what was the estimated mark in maths?

(5 marks)

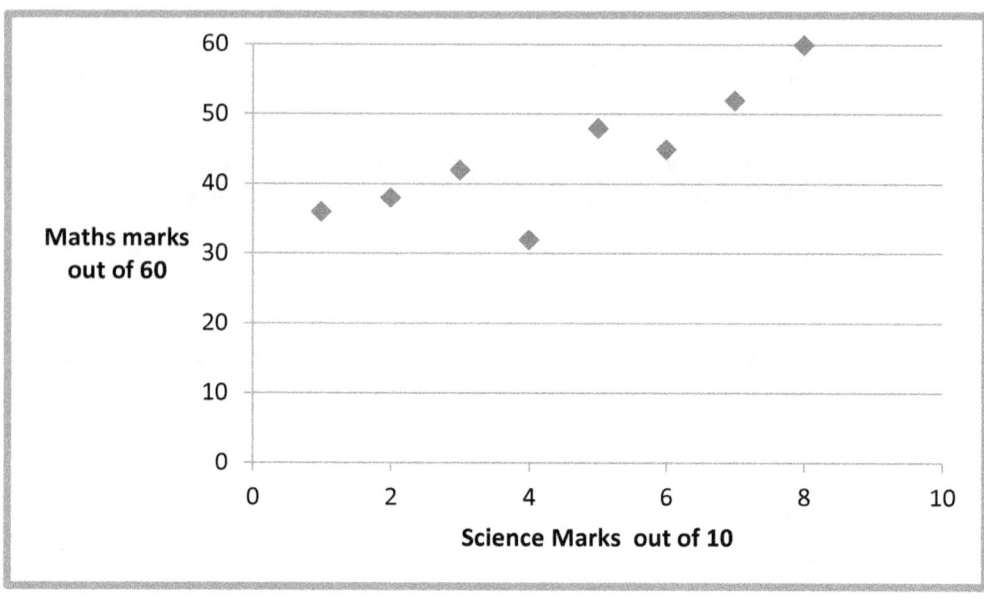

Answers to Practice Questions (non-calculator section)

(1) Answers: Mode = 2, Median = 2.5 and Range = 14
(2) Answer: 48 cm^2
(3) Answers:
    (i)   The angle at the centre = <u>2×</u> the angle subtended at the circumference
    (ii)   The radius meets a circle tangent at <u>90°</u>
    (iii)   The angle between the tangent and the chord = <u>the angle in the alternate segment</u>
    (iv)   The opposite angles of a cyclic quadrilateral add up to <u>180°</u>
    (v)   The triangle in a semi-circle has an angle of <u>90°</u> at the circumference
(4) Answer: $\frac{\sqrt{3}}{3}$
(5) Answers: (a) $(1\frac{1}{2}, 2)$ (b) $y = 2x - 1$
(6) Answer: $\frac{3}{4}$
(7) Answer: 130 times
(8) Answer: 18°

(9) Answer: $\dfrac{3}{36} = \dfrac{1}{12}$

(10) Answer: (a) The correlation in this particular test is **positive**

(b) Draw a 'line of best fit' between the points as shown **in the scatter graph below.**

(c) The **estimated mark in maths is around 50** (a mark between 45 -55 is acceptable)

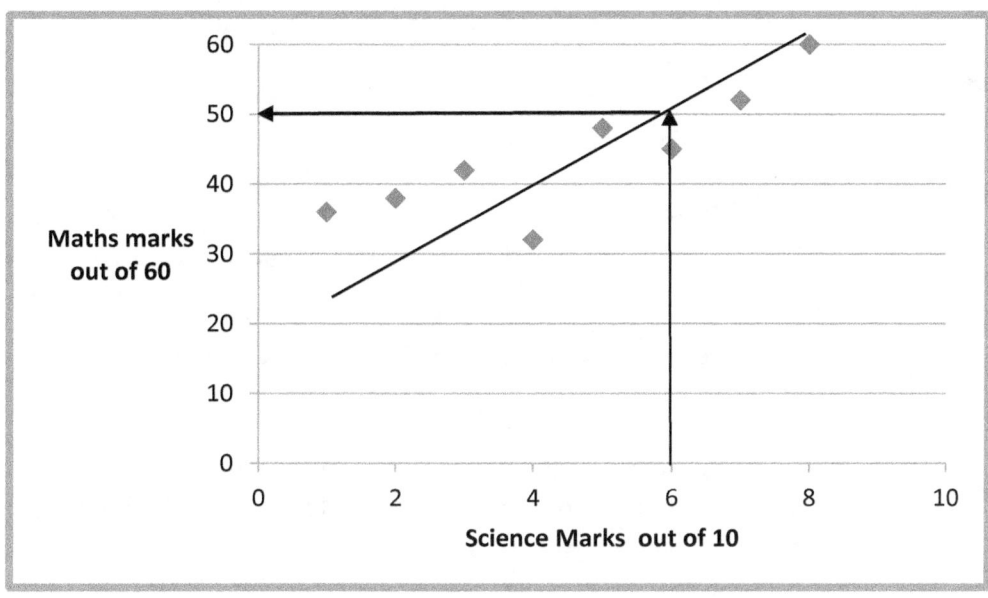

Practice Questions 6: Calculators Allowed

(1) Calculate the value of $\dfrac{2.15 \times 6.76 + 17.1 \times 1.4}{13.4 \times 2.2}$ to 5 decimal places.

(2 marks)

(2) If $f(x) = 3x^2 - 2x - 1$, find $f(x - 2)$ and then plot the equation $f(x - 2)$

(4 marks)

(3) In the triangle below find angle B (3 marks)

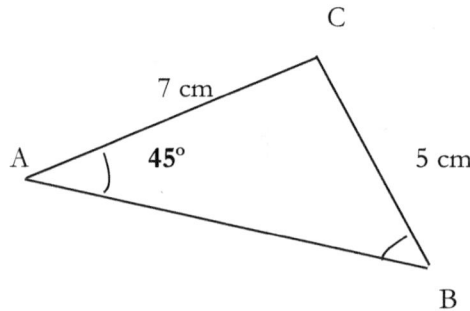

(4) Solve the simultaneous equations $2x + 3y = 1$ and $3x + 2y = 3$

(4 marks)

(5) Use the quadratic formula to solve the equation $y = 5x^2 + 2x - 1$. Give your answers to correct to 1 decimal place

(3 marks)

(6) In the graph below the equation $y = 3\sin(x) + 1$ is plotted. (i) What is the maximum and minimum value of y? (ii) When $x = 0$, what is the value of y?

(5 marks)

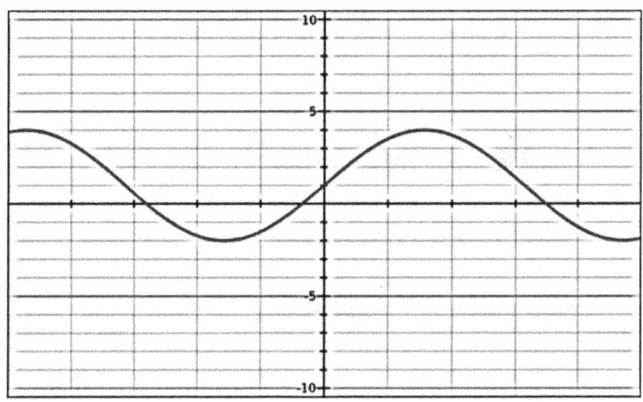

(7) Simplify the expression $\frac{2}{x-3} - \frac{3}{x+3} + \frac{1}{x^2-9}$ (5 marks)

(8) 40 year11 pupils were weighed. The data is shown in the grouped frequency table below. Find the estimated mean value of the weight correct to 1 decimal places. (4 marks)

Weight (w Kg)	Frequency (f)
70 < w ≤ 80	3
60 < w ≤ 70	22
50 < w ≤ 60	11
40 < w ≤ 50	4
30 < w ≤ 40	0

(9)(a) Find the nth term of the sequence 14, 11, 8, 5, ……. (b)Hence find the 50th term 4 marks

(10) ABC is a right angled triangle as shown below. Find the length of the side CB. You are given that AC = 6cm and AB =15cm 3 marks

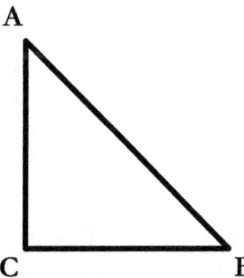

(11) Translate the points P (1, 3) and Q (2, 5) by the translation vector $\begin{bmatrix} -2 \\ 3 \end{bmatrix}$ and give the new co=ordinates of P and Q   2 marks

Answers: Practice – Calculator Paper

(1) Answer: 1.30509 to 5 decimal places

(2) Answer:(a) $f(x - 2) = 3x^2 - 14x + 15$

Answer: (b) See graph below:

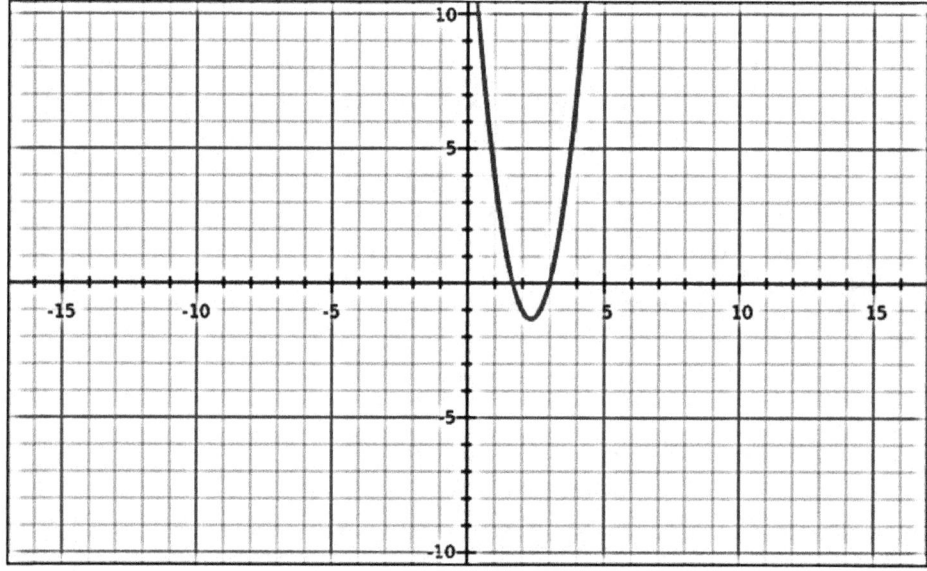

(3) Answer: B = 81.9°

(4) Answers: $x = \frac{7}{5}$ or $1\frac{2}{5}$ and $y = \frac{-3}{5}$

(5) Answer: x = 0.29, x = -0.69 (to two decimal places)

(6) Answers: (i) The maximum value of y is 4 and the minimum value of y is -2. (ii) When x = 0, y = 1

(7) Answer: $\dfrac{16-x}{x^2-9}$

(8) Answer: 61 kg

(9) Answer: (a) 17 – 3n (b) –133

(10) Answer: 13.75cm

(11) Answer: P'(-1, 6) and Q'(0, 8)

You are given a formula list at the beginning of each test paper. Typically these are:

The roots of a quadratic equation $ax^2 + bx + c$ are $x = \dfrac{-b \pm \sqrt{b^2 - 4ac}}{2a}$, where $a \neq 0$

The sine rule $\dfrac{a}{SinA} = \dfrac{b}{SinB} = \dfrac{c}{SinC}$

Cosine rule: $a^2 = b^2 + c^2 - 2bcCosA$

Area of triangle: $A = \dfrac{1}{2}abSinC$

Volume of sphere: $V = \dfrac{4}{3}\pi r^3$

Volume of cone: $V = \dfrac{1}{3}\pi r^2 h$

Curved surface of cone $= \pi rl$ (where l is the slanting length)

Volume of pyramid: $V = \dfrac{1}{3}Ah$

Area of Trapezium $= \dfrac{1}{2}(a + b)h$ (a and b are opposite parallel sides and h is the perpendicular height)

Volume of prism = Area of cross section × length

Note: In the actual exam, pass marks for grade boundaries are adjusted each year depending on the difficulty of the questions. As a rule of thumb around 35% in each paper should get you a grade C. However if you aim to get just over 45% in each paper you are very likely to get a Grade B. Between 60% – 75% will get you a grade A and 75% plus will get you an A*)

www.ingramcontent.com/pod-product-compliance
Lightning Source LLC
Chambersburg PA
CBHW050213230526
45470CB00001B/370